# Chemical Energy and Exergy

## An Introduction to Chemical Thermodynamics for Engineers

# Chemical Energy and Exergy

## An Introduction to Chemical Thermodynamics for Engineers


**Norio Sato**
Hokkaido University
NS Electrochemical Laboratory
Hokkaido, Japan



2004

ELSEVIER
Amsterdam - Boston - Heidelberg - London - New York - Oxford - Paris
San Diego - San Francisco - Singapore - Sydney - Tokyo

| | | | |
|---|---|---|---|
| **ELSEVIER B.V.** | ELSEVIER Inc. | ELSEVIER Ltd | ELSEVIER Ltd |
| **Sara Burgerhartstraat 25** | 525 B Street, Suite 1900 | The Boulevard, Langford Lane | 84 Theobalds Road |
| **P.O. Box 211, 1000 AE** | San Diego, CA 92101-4495 | Kidlington, Oxford OX5 1GB | London WC1X 8RR |
| **Amsterdam, The Netherlands** | USA | UK | UK |



First edition 2004

Library of Congress Cataloging in Publication Data
A catalog record is available from the Library of Congress.

British Library Cataloguing in Publication Data
A catalogue record is available from the British Library.

ISBN: 0-444-51645-X

Transferred to digital printing 2007

# PREFACE

This book is a beginner's introduction to chemical thermodynamics for engineers. According to the author's experience in teaching physical chemistry, chemical thermodynamics is the most difficult part for junior students to understand. Quite a number of students tend to lose their interest in the subject when the concept of entropy has been introduced in the lecture of chemical thermodynamics. Having had the practice of chemical technology after their graduation, however, they realize acutely the need of physical chemistry and begin studying chemical thermodynamics again.

The difficulty in learning chemical thermodynamics stems mainly from the fact that it appears too conceptual and much too complicated with many formulae. In this textbook efforts have been made to visualize as clearly as possible the main concepts of thermodynamic quantities such as enthalpy and entropy, thus making them more perceivable. Furthermore, intricate formulae in thermodynamics have been discussed as functionally unified sets of formulae to understand their meaning rather than to mathematically derive them in detail.

Most textbooks in chemical thermodynamics place the main focus on the equilibrium of chemical reactions. In this textbook, however, the affinity of irreversible processes, defined by the second law of thermodynamics, has been treated as the main subject. The concept of affinity is applicable in general not only to the processes of chemical reactions but also to all kinds of irreversible processes.

This textbook also includes electrochemical thermodynamics in which, instead of the classical phenomenological approach, molecular science provides an advanced understanding of the reactions of charged particles such as ions and electrons at the electrodes.

Recently, engineering thermodynamics has introduced a new thermodynamic potential called *exergy*, which essentially is related to the concept of the affinity of irreversible processes. This textbook discusses the relation between exergy and affinity and explains the *exergy balance diagram* and *exergy vector diagram* applicable to exergy analyses in chemical manufacturing processes.

This textbook is written in the hope that the readers understand in a broad way the fundamental concepts of energy and exergy from chemical thermodynamics in practical applications. Finishing this book, the readers may easily step forward further into an advanced text of their specified line.

The author finally expresses his deep gratitude to those who have contributed to the present state of chemical thermodynamics on which this book is based. He also thanks Mrs. Y. Sato for her assistance.

Norio Sato

Sapporo, Japan
December 2003

# CONTENTS

# CHAPTER 1

# THERMODYNAMIC STATE VARIABLES

Chemical thermodynamics deals with the physicochemical state of substances. All physical quantities corresponding to the macroscopic property of a physico-chemical system of substances, such as temperature, volume, and pressure, are thermodynamic variables of the state and are classified into intensive and extensive variables. Once a certain number of the thermodynamic variables have been specified, then all the properties of the system are fixed. This chapter introduces and discusses the characteristics of intensive and extensive variables to describe the physicochemical state of the system.

## 1. 1. Thermodynamic Systems.

In physics and chemistry we call an ensemble of substances a *thermodynamic system* consisting of atomic and molecular particles. The system is separated from the surroundings by a boundary interface. The system is called *isolated* when no transfer is allowed to occur of substances, heat, and work across the boundary interface of the system as shown in Fig. 1.1. The system is called *closed* when it allows both heat and work to transfer across the interface but is impermeable to substances. The system is called *open* if it is completely permeable to substances, heat, and work. The open system is the most general and it can be regarded as a part of a closed or isolated system. For instance, the universe is an isolated system, the earth is regarded as a closed system, and a creature such as a human being corresponds to an open system.

Ordinarily, the system may consist of several *phases*, whose interior in the state of equilibrium is homogeneous throughout its extent. The system, if composed for instance of only liquid water, is a single phase; and if made up for instance of liquid water and water vapor, it is a two phase system. The single phase system is frequently called a *homogeneous system*, and a multiphase system is called *heterogeneous*.

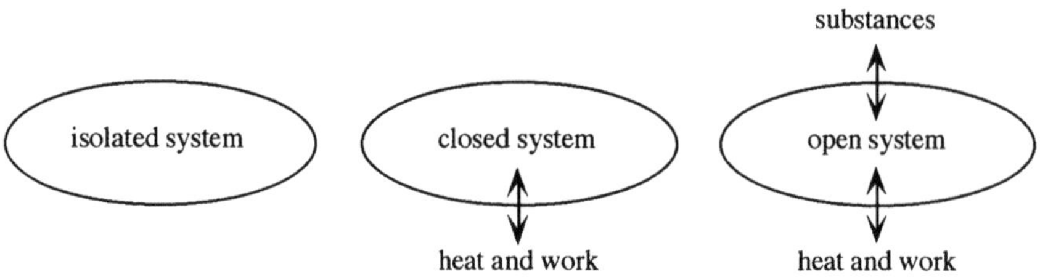


Fig. 1.1. Physicochemical systems of substance ensembles.

**1. 2. Variables of the State.**

All observable quantities of the macroscopic property of a thermodynamic system, such as the volume $V$, the pressure $p$, the temperature $T$, and the mass $m$ of the system, are called *variables of the state*, or *thermodynamic variables*. In a state of the system all observable variables have their specified values. In principle, once a certain number of variables of the state are specified, all the other variables can be derived from the specified variables. The state of a pure oxygen gas, for example, is determined if we specify two freely chosen variables such as temperature and pressure.

These minimum number of variables that determine the state of a system are called the *independent variables*, and all other variables which can be functions of the independent variables are *dependent variables* or *thermodynamic functions*. For a system where no external force fields exists such as an electric field, a magnetic field and a gravitational field, we normally choose as independent variables the combination of pressure-temperature-composition or volume-temperature-composition.

In chemistry we have traditionally expressed the amount of a substance $i$ in a system of substances in terms of the number of moles $n_i = m_i / M_i$ instead of its mass $m_i$, where $M_i$ denotes the gram molecular mass of the substance $i$. The composition of the system of substances is expressed accordingly by the *molar fraction* $x_i$ as defined in Eq. 1.1:

$$x_i = \frac{n_i}{\sum_i n_i} = \frac{n_i}{n}, \qquad \sum_i x_i = 1 . \tag{1.1}$$

In the case of solutions (liquid or solid mixtures), besides the molar fraction, we frequently use for expressing the solution composition the *molar concentration* (or *molarity*) $c_i$, the number of moles for unit volume of the solution, and the *molality* $m_i$, the number of moles for unit mass of the solvent (main component substance of the solution):

$$c_i = \frac{n_i}{V}\ \mathrm{mole \cdot m^{-3}}, \qquad m_i = \frac{n_i}{M_\mathrm{S}}\ \mathrm{mol \cdot kg^{-1}}, \tag{1.2}$$

where $V$ is the volume of 1 $m^3$ of the solution and $M_\mathrm{S}$ is the mass of 1 kg of the solvent.

## 1. 3. Extensive and Intensive Variables.

The variables whose values are proportional to the total quantity of substances in the system are called ***extensive variables*** or ***extensive properties***, such as the volume $V$ and the number of moles $n$. The extensive variables, in general, depend on the size or quantity of the system. The masses of parts of a system, for instance, sum up to the total mass of the system, and doubling the mass of the system at constant pressure and temperature results in doubling the volume of the system as shown in Fig. 1.2.

On the contrary, the variables that are independent of the size and quantity of the system are called *intensive variables* or *intensive properties*, such as the pressure $p$, the temperature $T$, and the mole fraction $x_i$ of a substance $i$. Their values are constant throughout the whole system in equilibrium and remain the same even if the size of the system is doubled as shown in Fig. 1.2.

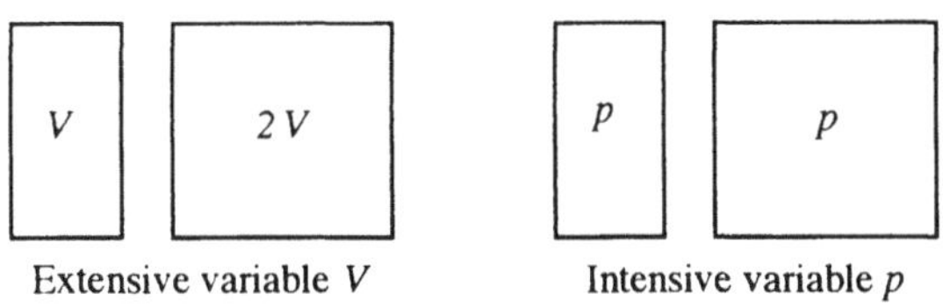


Fig. 1.2. Extensive and intensive variables in a physicochemical system.

## 1. 4. Partial Molar Quantities.

An extensive variable may be converted into an intensive variable by expressing it per one mole of a substance, namely, by partially differentiating it with respect to the number of moles of a substance in the system. This partial differential is called in chemical thermodynamics the *partial molar quantity*. For instance, the volume $v_i$ for one mole of a substance $i$ in a homogeneous mixture is given by the derivative (partial differential) of the total volume $V$ with respect to the number of moles of substance $i$ as shown in Eq. 1.3:

$$v_i = \left(\frac{\partial V}{\partial n_i}\right)_{T,\,p,\,n_j}, \tag{1.3}$$

where the subscripts $T, p$ and $n_j$ on the right hand side mean that the temperature $T$, pressure $p$, and all $n_j$'s other than $n_i$ are kept constant in the system. The derivative $v_i$ is the *partial molar volume* of substance $i$ at constant temperature and pressure and expresses the increase in volume that results from the addition of one mole of substance $i$ into the system whose initial volume is very large.

In general, the partial molar volume $v_i$ of substance $i$ in a homogeneous multiconstituent mixture differs from the molar volume $v_i^0 = V/n_i$ of the pure substance $i$. When we add one mole of pure substance $i$ into the mixture, its volume changes from the molar volume $v_i^0$ of the pure substance $i$ to the partial molar volume $v_i$ of substance $i$ in the mixture as shown in Fig. 1.3(a). In a system of a single substance, by contrast, the partial molar volume $v_i$ is obviously equal to the molar volume $v_i^0$ of the pure substance $i$.

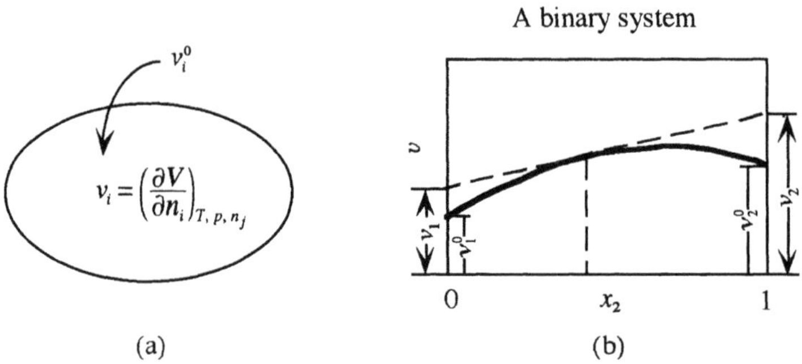


Fig. 1.3. Partial molar volume: (a) the molar volume $v_i^0$ of a pure substance $i$ and the partial molar volume $v_i$ of substance $i$ in a homogeneous mixture; (b) graphical determination of the partial molar volumes of constituent substances in a homogeneous binary system by the Bakhuis-Rooseboom Method: $v = V/(n_1 + n_2)$ = the mean molar volume of a binary mixture; $x_2$= the molar fraction of substance 2; $v_1 = v - x_2(\partial v/\partial x_2)$ = the partial molar volume of substance 1; $v_2 = v - (1 - x_2)(\partial v/\partial x_2)$ = the partial molar volume of substance 2. [Ref. 1.]

In a system of a homogeneous mixture containing multiple substances the total volume $V$ is given by the sum of the partial molar volumes of all the constituent substances each multiplied by the number of moles as shown in Eq. 1.4:

$$V = \Sigma_i n_i v_i. \tag{1.4}$$

The partial molar volume $v_i$ of a substance $i$ is of course not identical with the molar volume $v = V/\Sigma_i n_i$ of the mixture.

Considering that the volume $V$ of a system is a *homogeneous function of the first degree* in the variables $n_i$, $\left[\text{Euler's theorem}; f(kn_1, kn_2) \equiv kf(n_1, n_2)\right]$, we can write through differentiation of Eq. 1.4 with respect to $n_j$ at constant temperature and pressure the equation expressed by:

$$\sum_i n_i(\partial v_i/\partial n_i)_{T,p} = 0 . \tag{1.5}$$

For a homogeneous binary mixture consisting of substance 1 and substance 2, we then have Eq. 1.6:

$$n_1\left(\frac{\partial^2 V}{\partial n_1\,\partial n_2}\right)_{T,p}+n_2\left(\frac{\partial^2 V}{\partial n_2\,\partial n_2}\right)_{T,p}=0,\qquad x_1\left(\frac{\partial v_1}{\partial n_2}\right)_{T,p}+x_2\left(\frac{\partial v_2}{\partial n_2}\right)_{T,p}=0. \tag{1.6}$$

Furthermore, Eq. 1.6 gives Eq. 1.7:

$$x_1\left(\frac{\partial v_1}{\partial x_2}\right)_{T,p}+x_2\left(\frac{\partial v_2}{\partial x_2}\right)_{T,p}=0. \tag{1.7}$$

From the molar volume $v = V/(n_1+n_2) = (1-x_2)\,v_1 + x_2\,v_2$ and its derivative $(\partial v/\partial x_2)_{T,p} = (v_2 - v_1)$ multiplied by $x_2$, we obtain Eq. 1.8:

$$v_1 = v - x_2\left(\frac{\partial v}{\partial x_2}\right)_{T,p}. \tag{1.8}$$

This equation 1.8 can be used to estimate the partial molar volume of a constituent substance in a binary mixture from the observed curve of the molar volume $v$ against the molar fraction $x_2$ as shown in Fig. 1.3(b).

## 1. 5. The Extent of a Chemical Reaction.

Let us consider a chemical reaction that occurs in a closed system. According to the law of *the conservation of mass*, the total sum of the mass of all the chemical substances remains constant in the system whatever the chemical reactions taking place.

The chemical reaction may be expressed by a formula shown in Eq. 1.9:

$$\nu_1\,\mathrm{R}_1 + \nu_2\,\mathrm{R}_2 \rightarrow \nu_3\,\mathrm{P}_3 + \nu_4\,\mathrm{P}_4, \tag{1.9}$$

where $\mathrm{R}_1$ and $\mathrm{R}_2$ are the chemical species being consumed (reactants), $\mathrm{P}_3$ and $\mathrm{P}_4$ are the chemical species being produced (products), and $\nu_1 \cdots \nu_4$ are the *stoichiometrical coefficients* of the reactants and products in the reaction, respectively. The stoichiometrical coefficient is negative for the reactants and positive for the products. The conservation of mass in the reaction is expressed by Eq. 1.10:

$$\nu_3\,M_3 + \nu_4\,M_4 + \nu_1\,M_1 + \nu_2\,M_2 = 0,\qquad \Sigma_i\,\nu_i M_i = 0, \tag{1.10}$$

where $M_i$ denotes the relative molecular mass of species $i$.

We express the change in the number of moles $n_i$ of each species as follows:

$$n_1 - n_1^0 = \nu_1\,\xi,\quad n_2 - n_2^0 = \nu_2\,\xi,\quad n_3 - n_3^0 = \nu_3\,\xi,\quad n_4 - n_4^0 = \nu_4\,\xi, \tag{1.11}$$

where $n_1^0 \cdots n_4^0$ denote the initial number of moles of the reaction species at the beginning of the reaction. The symbol $\xi$ represents the degree of advancement of the reaction. In chemical thermodynamics it is called the *extent of reaction*.

The initial state of a reaction is defined by $\xi = 0$, and the state at which $\xi = 1$ corresponds to the final state where all the reactants ($\nu_1$ moles of $R_1$ and $\nu_2$ moles of $R_2$) have been converted to the products ($\nu_3$ moles of $P_3$ and $\nu_4$ moles of $P_4$) as shown in Fig. 1.4. We say *one equivalent of reaction* has occurred when a system undergoes a chemical reaction from the state of $\xi = 0$ to the state of $\xi = 1$.

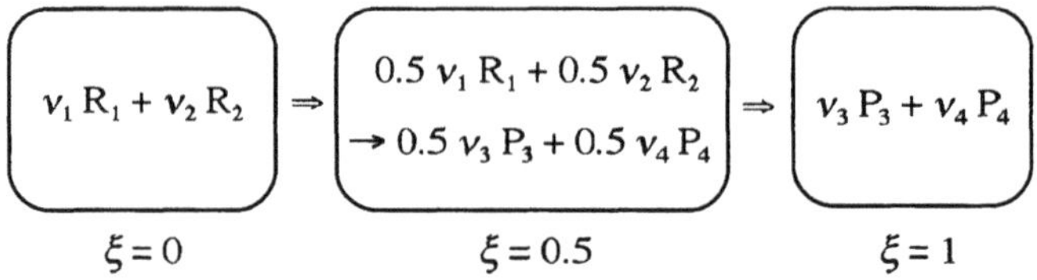


Fig. 1.4. The extent of a chemical reaction.

Equation 1.11 gives us the differential of the extent of reaction $d\xi$ shown in Eq. 1.12:

$$\frac{dn_1}{\nu_1} = \frac{dn_2}{\nu_2} = \frac{dn_3}{\nu_3} = \frac{dn_4}{\nu_4} = d\xi \tag{1.12}$$

To take an instance, we consider the following two reactions in a system consisting of a solid phase of carbon and a gas phase containing molecular oxygen, carbon monoxide and carbon dioxide:

$$2\,C_{(solid)} + O_{2\,(gas)} \rightarrow 2\,CO_{(gas)}, \qquad \text{Reaction 1,}$$

$$C_{(solid)} + O_{2\,(gas)} \rightarrow CO_{2\,(gas)}, \qquad \text{Reaction 2.}$$

For these two reactions the following equations hold between the extents of reactions $\xi$ and the number of moles of reaction species $n_i$:

$$dn_C = -2\,d\xi_1 - d\xi_2\,, \quad dn_{O_2} = -\,d\xi_1 - d\xi_2, \quad dn_{CO} = 2\,d\xi_1, \quad dn_{CO_2} = d\xi_2\,.$$

The reaction rate v is expressed by the differential of the extent of reaction $\xi(t)$ with respect to time $t$ as shown in Eq. 1.13:

$$\mathrm{v} = \frac{d\xi(t)}{dt}. \tag{1.13}$$

The *reaction rate* may also be expressed by the time-differential of the mass or the number of moles of reaction species. For a single reaction the reaction rate in terms of the extent of reaction is related with the reaction rate in terms of the mass $m_i$ or the number of moles $n_i$ of reaction species as shown in Eq. 1.14:

$$\frac{dn_i}{dt} = \nu_i \, v, \qquad \frac{dm_i}{dt} = \nu_i M_i \, v. \tag{1.14}$$

The extent of reaction is an extensive property, and it can apply not only to chemical reactions but also as the extent of change to all physicochemical processes such as diffusion, melting, boiling, and solid state transformation.

# CHAPTER 2

# CONSERVATION OF ENERGY

The first law of thermodynamics provides the concept of energy, which is defined based on empirical knowledge as a physical quantity of the state of thermodynamic systems. In reality energy presents itself in various forms such as thermal, mechanical, chemical, electrical, magnetic, photonic energy, etc. These various forms of energy can be converted into one another with some restriction in thermal energy. The first law also expresses the empirical principal that the total amount of energy is conserved whatever energy conversion may take place. Moreover, thermodynamics introduces two energy functions called the internal energy and the enthalpy depending on the choice of independent variables. This chapter discusses the characteristics of these two energy functions.

## 2. 1. Energy as a Physical Quantity of the State.

Thermodynamics has provided in its first law the concept of *energy*, which is a self-evident quantity empirically defined for the capacity that a thermodynamic system possesses of doing physicochemical work (energy = en+erg). The *first law of thermodynamics* indicates that the energy of an isolated system is constant and that the change in the energy of a closed system is equal to the amount of energy received from or released out of the system (the principal of the conservation of energy). Energy is an extensive property and its recommended SI unit is joule J whose dimension is $kg\cdot m^2\cdot s^{-2}$.

Energy may be classified into varieties such as mechanical, thermal, chemical, photonic, electric, and magnetic energy. These different forms of energy, however, can theoretically be converted one to one in each other, except for thermal energy whose conversion is restricted by the second law of thermodynamics as will be mentioned in the following chapter. If the system undergoes nuclear reactions, the mass of substances converts into what is called the nuclear energy. We won't discuss nuclear reactions in this book, however.

In general, mechanical energy or work is expressed by the product of the force $f$ affecting a body and the distance $\Delta l$ over which the body moves in the direction of the force: $f \cdot \Delta l$. A change in the volume of a system causes mechanical work done by the system or performed on the system, whose magnitude corresponds to the product of the pressure $p$ and the volume change $\Delta V$: $p \cdot \Delta V$. Further, electric energy is represented by the product of the voltage and the electric charge. Furthermore, thermal energy reversibly received by a system equals the product of the absolute temperature $T$ and the change in thermal entropy $\Delta S$ in the system:entropy will be described in the following chapter. We may hence conceptually assume the following relation in Eq. 2.1:

$$\text{Energy} = \text{Intensive variable} \times \text{Conjugate extensive variable}. \tag{2.1}$$

where energy is formally expressed by the product of conjugated intensive and extensive variables.

## 2. 2. Conservation of Energy.

Let us consider a closed system which can exchange heat and work but not substances with its surroundings. The exchange of heat and work takes place through the boundary interface of the system. The energy of the system then increases by an amount equal to the heat and work absorbed from the surroundings. We define the *internal energy* $U$ of the system as a state property whose infinitesimal change $dU$ is equal to the sum of infinitesimal heat $dQ$ and infinitesimal work $dW$ received by the system as shown in Eq. 2.2:

$$dU = dQ + dW, \tag{2.2}$$

where the heat and work received by the system are positive quantities, while those released out of the system are negative as shown in Fig. 2.1. The integral of internal energy $\int dU = \int dQ + \int dW$ from a certain initial state to a certain final state of the system is always independent of the route followed, though each of $\int dQ$ and $\int dW$ may depend on the rout. The *internal energy* is hence defined as a state property. We also call the heat $dQ$ and the work $dW$ the *energy transferred* across the boundary between the system and the surroundings. Internal energy, heat, and work must of course be measured in the same unit.

Work can have different forms such as compression-expansion-, electric-, magnetic work and other forms. The amount of work done by these different forms can be measured in the same scale of joule that we normally use for measuring heat and energy. Work, heat, and internal energy thus present themselves in the same category of energy. Thermodynamics however shows us that the heat differs somehow in its quality from the other forms of energy in that the energy of heat (thermal energy) can not be completely converted one to one into the other forms of energy as will be discussed in the following chapter.

If the work done by the system is only due to a change in volume of the system under the pressure $p$, we obtain $dW = -p\,dV$. Then, Eq. 2.2 yields Eq. 2.3:

$$dU = dQ - p\,dV, \tag{2.3}$$

where $p$ is the internal pressure of the system. In thermodynamics we usually assume an ideal process called *reversible* in which all changes take place in quasi-equilibrium. The external pressure then is equal to the internal pressure of the system. We thus assume for the reversible process that the pressure $p$ in Eq. 2.3 is equal to the internal pressure of the system itself.

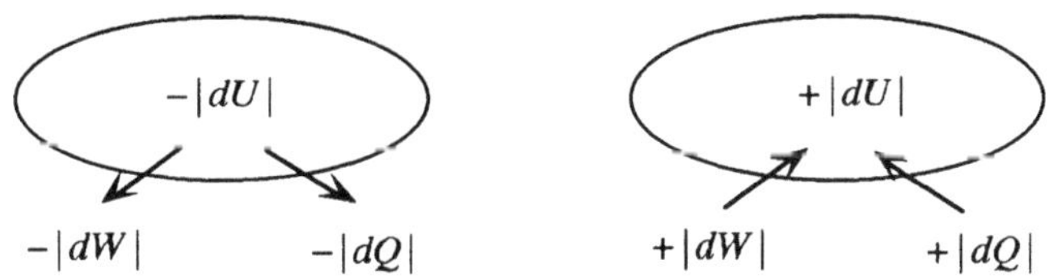


Fig. 2.1. Conservation of energy in a closed system.

**2. 3. Internal Energy $U$ with Independent Variables $T$, $V$, and $\xi$.**

We now consider a homogeneous closed system containing $c$ species of substances in which *a chemical reaction* occurs in a reversible way. The internal energy, $U$, is a function of the state of the system, and hence may be expressed in terms of the independent variables that characterize the state. If the state of the system is determined by the independent variables temperature $T$, volume $V$, and extent of reaction $\xi$ as shown in Fig. 2.2, we have $U = U(T, V, n_1^0 \cdots n_c^0, \xi)$, where $n_1^0 \cdots n_c^0$ are the initial number of moles of the species of substances. The total differential of the internal energy $U$ is then given by Eq. 2.4:

$$dU = \left(\frac{\partial U}{\partial T}\right)_{V,\xi} dT + \left(\frac{\partial U}{\partial V}\right)_{T,\xi} dV + \left(\frac{\partial U}{\partial \xi}\right)_{T,V} d\xi. \tag{2.4}$$

From Eqs. 2.3 and 2.4 we obtain Eq. 2.5 for transferred heat $dQ$:

$$dQ = \left(\frac{\partial U}{\partial T}\right)_{V,\xi} dT + \left\{\left(\frac{\partial U}{\partial V}\right)_{T,\xi} + p\right\} dV + \left(\frac{\partial U}{\partial \xi}\right)_{T,V} d\xi. \tag{2.5}$$

Equation 2.5 can also be expressed by Eq. 2.6:

$$dQ = C_{V,\xi}\,dT + l_{T,\xi}\,dV + u_{T,V}\,d\xi, \tag{2.6}$$

where $C_{V,\xi}$, $l_{T,\xi}$, and $u_{T,V}$ are the *thermal coefficients* for the variables $T$, $V$, and $\xi$.

The coefficient, $C_{V,\xi} = (\partial Q/\partial T)_{V,\xi} = (\partial U/\partial T)_{V,\xi}$, is the amount of heat required to raise the temperature of the system by unit degree at constant $V$ and $\xi$; it is called the *heat capacity* of the system at constant volume and composition. The coefficient, $l_{T,\xi} = (\partial Q/\partial V)_{T,\xi} = \{(\partial U/\partial V)_{T,\xi} + p\}$, is the heat that must be supplied to the system for unit increase in volume at constant temperature, and may be called the *latent heat of volume change* of the system. For an ideal gas, whose internal energy is independent of the volume $(\partial U/\partial V)_{T,\xi} = 0$, we have $l_{T,\xi} = p$.

The coefficient of $u_{T,V} = (\partial Q/\partial \xi)_{T,V} = (\partial U/\partial \xi)_{T,V}$ is the heat received by the system when the reaction proceeds by an extent of reaction $d\xi$ at constant temperature and volume, and its integral from $\xi = 0$ to $\xi = 1$ is the *heat of reaction* at constant volume and temperature, $Q_{T,V}$:

$$Q_{T,V} = \int_0^1 u_{T,V}\, d\xi. \tag{2.7}$$

In particular, if $u_{T,V}$ is independent of $\xi$, $Q_{T,V}$ is given by $Q_{T,V} = u_{T,V}(\xi_1 - \xi_0)$, and for one equivalent extent of reaction ($\xi_1 - \xi_0 = 1$) we obtain the heat of reaction $Q_{T,V} = u_{T,V}$ at constant volume.

The reaction is called *exothermic* if the heat of reaction is negative; whereas, the reaction is *endothermic* if it is positive.

Fig. 2.2. Thermodynamic energy functions: (a) Internal energy $U$, (b) Enthalpy $H$.

## 2. 4. Enthalpy *H* with Independent Variables *T*, *p*, and $\xi$.

If we choose $T$, $p$, and $\xi$ as independent variables, the total differential of $U$ is given by Eq. 2.8:

$$dU = \left(\frac{\partial U}{\partial T}\right)_{p,\xi} dT + \left(\frac{\partial U}{\partial p}\right)_{T,\xi} dp + \left(\frac{\partial U}{\partial \xi}\right)_{T,p} d\xi. \tag{2.8}$$

Volume $V$ is no longer an independent variable but a function of $T$, $p$, and $\xi$: $V(T,p,\xi)$. The total differential $dV(T,p,\xi)$ in Eq. 2.3 can then be expressed by Eq. 2.9:

$$dV = \left(\frac{\partial V}{\partial T}\right)_{p,\xi} dT + \left(\frac{\partial V}{\partial p}\right)_{T,\xi} dp + \left(\frac{\partial V}{\partial \xi}\right)_{T,p} d\xi. \tag{2.9}$$

By writing $dQ$ from Eq. 2.3 explicitly and using Eqs. 2.8 and 2.9, we thus obtain Eq. 2.10:

$$dQ = \left\{\left(\frac{\partial U}{\partial T}\right)_{p,\xi} + p\left(\frac{\partial V}{\partial T}\right)_{p,\xi}\right\} dT + \left\{\left(\frac{\partial U}{\partial p}\right)_{T,\xi} + p\left(\frac{\partial V}{\partial p}\right)_{T,\xi}\right\} dp$$
$$+ \left\{\left(\frac{\partial U}{\partial \xi}\right)_{T,p} + p\left(\frac{\partial V}{\partial \xi}\right)_{T,p}\right\} d\xi. \tag{2.10}$$

We realize in Eq. 2.10 that, for the independent variables $T$, $p$, and $\xi$, it is advantageous to use the thermodynamic energy function $H$ called *enthalpy* as defined in Eq. 2.11:

$$H = U + p\,V, \tag{2.11}$$

which may also be called the *heat content* or *heat function* in the field of engineering thermodynamics. The word of enthalpy means "heating up" in Greek.

Using this energy function $H$, we obtain from Eq. 2.3 the expression of the heat received by the system as shown in Eq. 2.12:

$$dQ = dH - p\,dV - V\,dp + p\,dV = dH - V\,dp, \tag{2.12}$$

which then yields Eq. 2.13:

$$dQ = \left(\frac{\partial H}{\partial T}\right)_{p,\xi} dT + \left\{\left(\frac{\partial H}{\partial p}\right)_{T,\xi} - V\right\} dp + \left(\frac{\partial H}{\partial \xi}\right)_{T,p} d\xi. \tag{2. 13}$$

Equation 2.13 may be expressed as follows:

$$dQ = C_{p,\xi}\,dT + h_{T,\xi}\,dp + h_{T,p}\,d\xi, \tag{2.14}$$

where $C_{p,\xi}$, $h_{T,\xi}$, and $h_{T,p}$ are the thermal coefficients for the variables $T$, $p$, and $\xi$. Comparing Eq. 2.13 with Eq. 2.14, we realize that; $C_{p,\xi} = (\partial H/\partial T)_{p,\xi}$ is the *heat capacity* of the system at constant pressure and composition; $h_{T,\xi} = \{(\partial H/\partial p)_{T,\xi} - V\}$ may be called the *latent heat of pressure change*, and $h_{T,p} = (\partial H/\partial \xi)_{T,p}$ is the *heat of reaction* at constant pressure and temperature:

$$C_{p,\xi} = \left(\frac{\partial H}{\partial T}\right)_{p,\xi}, \quad h_{T,\xi} = \left(\frac{\partial H}{\partial p}\right)_{T,\xi} - V, \quad h_{T,p} = \left(\frac{\partial H}{\partial \xi}\right)_{T,p}. \tag{2.15}$$

The heat capacity $C_{p,\xi}$ is an extensive property and, for a mixture of substances $i$, is given as the sum of the partial molar heat capacity $c_{p,i}$ of all the constituent substances each multiplied by the number of moles $n_i$ of $i$ as shown in Eq. 2.16:

$$C_{p,\xi} = \sum_i n_i\,c_{p,i}, \qquad c_{p,i} = \left(\frac{\partial C_{p,\xi}}{\partial n_i}\right)_{T,p\,n_j}. \tag{2.16}$$

The latent heat of pressure change $h_{T,\xi}$, which is usually negative, is the amount of heat that must be removed from the system for unit increase in pressure to maintain constant temperature when the system is compressed at constant composition. For an ideal gas in which $pV = nRT$ and $(\partial U/\partial p)_{T,\xi} = 0$, the second term on the right hand side of Eq. 2.10 gives us $h_{T,\xi} = (\partial U/\partial p)_{T,\xi} + p\,(\partial V/\partial p)_{T,\xi}$. We then obtain the latent heat of pressure change as shown in Eq. 2.17:

$$h_{T,\xi} = \left(\frac{\partial U}{\partial p}\right)_{T,\xi} + p\left(\frac{\partial V}{\partial p}\right)_{T,\xi} = -\frac{n\,RT}{p} = -V, \qquad \text{ideal gas,} \tag{2.17}$$

indicating that for an ideal gas $h_{T,\xi}$ equals $-V$. From Eq. 2.15 we thus have the enthalpy of an ideal gas as follows:

$$\left(\frac{\partial H}{\partial p}\right)_{T,\xi} = 0; \qquad \text{ideal gas,} \tag{2.18}$$

which indicates that the enthalpy of an ideal gas is independent of the pressure of the gas.

The coefficient $h_{T,p} = (\partial H/\partial \xi)_{T,p}$ is the differential of the amount of heat that must be added to or extracted from the system for unit change in the extent of reaction at constant $p$ and $T$, and its integral from $\xi = 0$ to $\xi = 1$ is the *heat of reaction* at constant pressure and temperature:

$$Q_{T,p} = \int_0^1 h_{T,p}\, d\xi. \tag{2.19}$$

If $h_{T,p}$ is independent of $\xi$, the heat of reaction $Q_{T,p}$ then is equal to $h_{T,p}$.

Figure 2.3 shows the relation between enthalpy $H$ and each of the variables of $T$, $p$, and $\xi$ for an ideal gas reaction, in which we assume that the heat of reaction is constant irrespective of the extent of reaction.

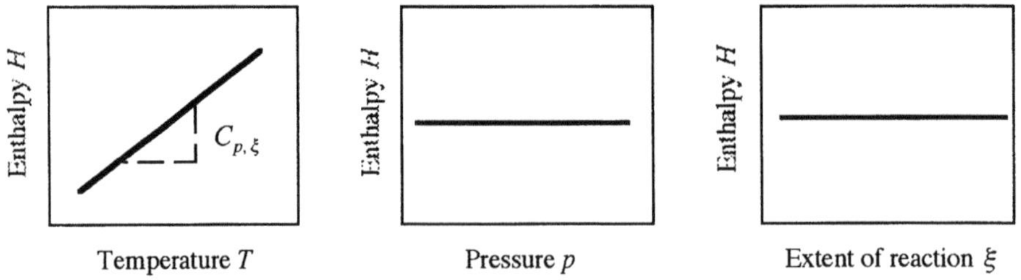


Fig. 2.3. Enthalpy as a function of temperature, pressure, and extent of reaction for an ideal gas reaction.

From Eqs. 2.5, 2.6, 2.10, 2.13 and 2.14 we obtain the following three equations 2.20, 2.21 and 2.22, which show the relationship between the thermal coefficients $C_{p,\xi}$, $h_{T,\xi}$, and $h_{T,p}$ for the variables $T$, $p$, and $\xi$, and the thermal coefficients $C_{V,\xi}$, $l_{T,\xi}$, and $u_{T,p}$ for the variables $T$, $V$, and $\xi$:

$$C_{p,\xi} = C_{V,\xi} + l_{T,\xi}\left(\frac{\partial V}{\partial T}\right)_{p,\xi}, \tag{2.20}$$

$$h_{T,\xi} = l_{T,\xi}\left(\frac{\partial V}{\partial p}\right)_{T,\xi}, \tag{2.21}$$

$$h_{T,p} = u_{T,V} + l_{T,\xi}\left(\frac{\partial V}{\partial \xi}\right)_{p,T}. \tag{2.22}$$

If we take, as an example, a closed system of a mixture of ideal gases in which a chemical reaction is occurring, then we have Eq. 2.23:

$$\left(\frac{\partial V}{\partial T}\right)_{p,\xi} = \frac{n R}{p}, \qquad \left(\frac{\partial V}{\partial \xi}\right)_{p,T} = \frac{R T}{p}\left(\frac{\partial n}{\partial \xi}\right) = \frac{\nu R T}{p}, \tag{2.23}$$

where $\nu = \Sigma_i \nu_i$ is the sum of stoichiometrical coefficients in the reaction. Furthermore, since $l_{T,\xi} = p$ for ideal gases as described in the foregoing in connection with Eq. 2.6, we obtain Eqs. 2.24 and 2.25 from Eqs. 2.20, 2.21, and 2.22:

$$C_{p,\xi} - C_{V,\xi} = n R, \tag{2.24}$$

$$h_{T,p} - u_{T,V} = \nu R T, \qquad \left(\frac{\partial H}{\partial \xi}\right)_{T,p} - \left(\frac{\partial U}{\partial \xi}\right)_{T,V} = \nu R T. \tag{2.25}$$

Thus for a gas reaction such as $N_{2\,(gas)} + 3H_{2\,(gas)} = 2NH_{3\,(gas)}$ for which $\nu = -2$, we obtain $(\partial H/\partial \xi)_{T,p} - (\partial U/\partial \xi)_{T,V} = -2\,RT$. This shows the relationship between the heat of the reaction at constant volume and that at constant pressure.

## 2. 5. Enthalpy and Heat of Reaction.

To describe the energy of a physicochemical system in which a chemical reaction takes place, it is convenient to make use of the internal energy $U$ if the reaction proceeds at constant volume or the enthalpy $H$ if the reaction proceeds at constant pressure. The system at constant volume undergoes no mechanical work and hence the change in internal energy is equal to the heat of the reaction. The system at constant pressure, in contrast, can receive work from or give off work to the surroundings as it changes its volume, so that the heat of reaction is not equivalent to the change in internal energy $U$ but to the change in enthalpy $H = U + pV$ of the system.

The heat of a reaction at constant temperature and pressure is normally defined as the change in enthalpy of the reaction system when the reactants are completely transformed into the products. The heat of a reaction, $(\partial H/\partial \xi)_{T,p}$, can thus be expressed in terms of the partial molar enthalpy, $h_i$, of reaction species $i$ given by Eq. 2.26 as shown in Eq. 2.27:

$$h_i = \left(\frac{\partial H}{\partial n_i}\right)_{T,p,j}, \tag{2.26}$$

$$\left(\frac{\partial H}{\partial \xi}\right)_{T,p} = \sum_i \left(\frac{\partial H}{\partial n_i}\right)_{T,p,j} \frac{dn_i}{\partial \xi} = \sum_i \nu_i h_i, \tag{2.27}$$

where $\nu_i$ is the stoichiometrical coefficient of substance $i$ in the reaction. From Eq. 2.27 we obtain, as an example, Eq. 2.28 for the heat of reaction for the formation of a compound AB from its constituent elements A and B, such as $S_{(solid)} + O_{2\,(gas)} \rightarrow SO_{2\,(gas)}$:

$$\left(\frac{\partial H}{\partial \xi}\right)_{T,p} = h_{AB} - \left(h_A + h_B\right) = h_{AB}^f, \tag{2.28}$$

where $h_{AB}^f$ represents the heat of the formation of compound AB at constant $p$ and $T$.

Recalling $\partial(\partial H/\partial \xi)/\partial T = \partial(\partial H/\partial T)/\partial \xi$, we have from Eq. 2.15 the heat of reaction at constant pressure as a function of the heat capacities, $C_p$, of all the reaction species. The temperature dependence of the heat of reaction at constant pressure is thus determined by the partial molar heat capacities, $c_{p,i}$, of the reaction species as shown in Eq. 2.29:

$$\frac{\partial}{\partial T}\left(\frac{\partial H}{\partial \xi}\right)_{T,p} = \left(\frac{\partial C_{p,\xi}}{\partial \xi}\right) = \sum_i \nu_i c_{p,i}. \tag{2.29}$$

This equation enables us to calculate the heat of a reaction at any temperature, provided that we know the value of the heat of the reaction at a specified temperature and that we know the partial molar heat capacities $c_{p,i}$ of all the species taking part in the reaction: $c_{p,i}$ may be equated to the molar heat capacities of the pure species in the case of gas reactions. By integrating Eq. 2.29 with respect to temperature we obtain Eq. 2.30 for the temperature dependence of the heat of reaction:

$$\left(\frac{\partial H}{\partial \xi}\right)_{T_2,p} - \left(\frac{\partial H}{\partial \xi}\right)_{T_1,p} = h_{T_2,p} - h_{T_1,p} = \int_{T_1}^{T_2} \sum_i \nu_i c_{p,i}\, dT. \tag{2.30}$$

This equation is used for estimating the heat of a reaction $(\partial H / \partial \xi)_{T_2,p}$ at a temperature $T_2$ when we know the value of the heat of the reaction $(\partial H / \partial \xi)_{T_1,p}$ at a specified temperature $T_1$ and the partial molar heat capacities $c_{p,i}$ of the reactants and products.

## 2. 6. Enthalpy of Pure Substances.

We now examine the enthalpy of a pure substance. Equation 2.15 shows that the enthalpy of a pure substance $i$ is a function of temperature $T$ and pressure $p$. A pure substance $i$ increases its enthalpy $H$ when it absorbs heat $Q$ at constant pressure. The differential of the molar enthalpy $dh_i$ is equivalent to the heat absorbed, $dq = dQ/dn_i$, for one mole of $i$ at constant pressure, and hence can be expressed in terms of the molar heat capacity $c_{p,i}$. The molar enthalpy also depends on the pressure of the system. The general equation to estimate the molar enthalpy of a substance can be derived from Eqs. 2.15 and 3.37, and we obtain Eq. 2.31:

$$h = h(0,0) + \int_0^T c_p(T, 0)\, dT + \int_0^p \left(\frac{\partial H}{\partial p}\right)_T dp\,, \tag{2.31}$$

where $h(0,0)$ is the enthalpy extrapolated to $p = 0$ and $T = 0$, $c_p(T, 0)$ is the heat capacity extrapolated to $p = 0$ at temperature $T$. If the substance undergoes any phase transformations in the temperature range concerned, the thermal and other energy changes associated with the phase transformations have to be taken into account.

In this equation 2.31 the second term on the right hand side is the thermal part and the third term is the pressure-dependent part. Normally, the pressure-dependent part is very small compared with the thermal part as shown in Eq. 2.18 for ideal gases, in which $(\partial H/\partial p)_T = 0$, and Eq. 7.27 for liquids and solids. For most purposes then the enthalpy may be regarded as independent of pressure and is given by Eq. 2.32

$$h = h(0,0) + \int_0^T c_p(T, 0)\, dT, \tag{2.32}$$

The enthalpy of a chemical substance at the standard state (298 K, 101.3 kPa) is called the standard enthalpy. In chemical thermodynamics, the standard enthalpy values of chemical elements in their stable states are all set zero, and hence the standard enthalpy of a chemical compound is represented by the heat of formation of the compound from its constituent elements at the standard state. Numerical values of the standard enthalpy of various chemical compounds thus obtained are tabulated in handbooks of chemistry.

# CHAPTER 3

# ENTROPY AS A STATE PROPERTY

The second law of thermodynamics provides a physical state property called *entropy* as an extensive variable relating to the capacity of energy distribution over the constituent particles in a physicochemical system. Also provided are two state properties called *free energy* (Helmholtz energy) and *free enthalpy* (Gibbs energy) both representing the available energy that the system possesses for physicochemical processes to occur in itself. This chapter discusses the creation of entropy due to the advancement of an irreversible process in a system, and elucidates the change in entropy caused by heat transfer, gas expansion, and mixing of substances. Also discussed is the affinity thermodynamically defined as the driving force of an irreversible process.

## 3. 1. Introduction to Entropy.

The energy of a physicochemical system is dependent on the substances that make the system. The substances, though macroscopically forming phases, are microscopically comprised of particles such as atoms, ions, and molecules constituting a particle ensemble. The energy of the system is distributed among individual particles in the ensemble, and the energy distribution over the constituent particles plays an important role in determining the property of the physicochemical system.

The second law of thermodynamics defines a state property called *entropy* as an extensive variable relating to the capacity of energy distribution over the constituent particles. The name of entropy comes from Greek meaning "progress or development". The energy of a system is not uniformly shared among the individual constituent particles but unevenly generating high and low energy particles. The distribution of energy among atomic and molecular particles is known to obey the Boltzmann statistics, which gives the most probable number of particles, $N_{\varepsilon_i}$, at an energy $\varepsilon_i$ in Eq. 3.1:

$$\frac{N_{\varepsilon_i}}{N}=\frac{e^{-\frac{\varepsilon_i}{kT}}}{\sum e^{-\frac{\varepsilon_i}{kT}}}, \tag{3.1}$$

where $N=\Sigma_i N_{\varepsilon_i}$ is the total number of particles, $k$ is the Boltzmann constant, $T$ is the absolute temperature. The exponential factor, $e^{-\varepsilon_i/kT}$, on the right hand side of Eq. 3.1 is well-known as the *Boltzmann factor*.

The denominator of the right hand side of Eq. 3.1 is relevant to the total number of the microscopic energy states of the system and is called the *particle partition function* $z$:

$$z=\sum e^{-\frac{\varepsilon_i}{kT}}. \tag{3.2}$$

Eqs. 3.1 and 3.2 give us an expression for the average internal energy $U/N$ of a particle in the system as shown in Eq. 3.3:

$$\frac{U}{N}=\frac{\sum \varepsilon_i N_{\varepsilon_i}}{\sum N_{\varepsilon_i}}=\frac{\sum \varepsilon_i e^{-\frac{\varepsilon_i}{kT}}}{\sum e^{-\frac{\varepsilon_i}{kT}}}=kT^2\left(\frac{\partial \ln z}{\partial T}\right)_{V,N}. \tag{3.3}$$

Statistical thermodynamics has defined, in addition to the particle partition function $z$, the *canonical ensemble partition function* $Z$ as follows:

$$Z=\sum e^{-\frac{U_i}{kT}}, \tag{3.4}$$

where $U_i$ is one of the allowed amounts of energy for a component system of the canonical system ensemble. The average internal energy $U$ of the ensemble is then obtained in the form similar to Eq. 3.3 as shown in Eq. 3.5:

$$U=kT^2\left(\frac{\partial \ln Z}{\partial T}\right)_{V,N}. \tag{3.5}$$

For a system consisting of the total number of particles $N$ and maintaining its total energy $U$ and volume $V$ constant, statistical thermodynamics defines the *entropy*, $S$, in terms of the logarithm of the total number of microscopic energy distribution states $\Omega(N,V,U)$ in the system as shown in Eq. 3.6:

$$S=k\ln \Omega\,(N,V,U). \tag{3.6}$$

The number of microscopic energy distribution states $\Omega(N,V,U)$ in the system is also related with the ensemble partition function $Z$. According to statistical mechanics, the entropy $S$ has been connected with the ensemble partition function $Z$ in the form of Eq. 3.7:

$$dS=k\,d\ln \Omega=k\,d\left(\ln Z+\frac{U}{kT}\right), \qquad S=k\ln Z+\frac{U}{T}+\text{constant}, \tag{3.7}$$

where the *absolute temperature* $T$ is defined by the second law of thermodynamics (thermodynamic temperature scale, Kelvin's temperature). Equation 3.7 gives us the unit of the entropy to be $J \cdot K^{-1}$. The entropy is obviously one of the extensive variables to specify the state of the system.

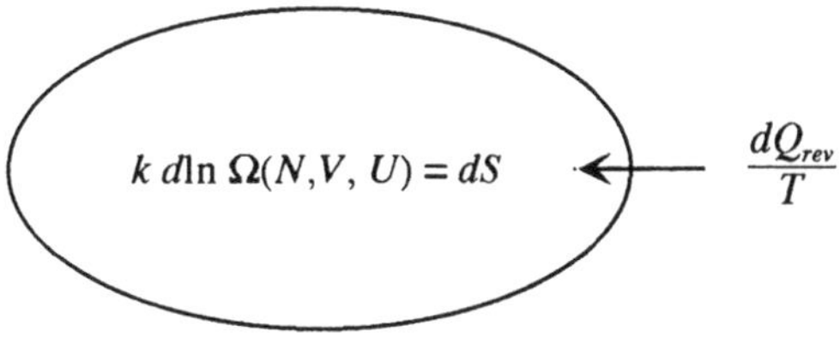


Fig. 3.1. Entropy change due to a reversible transfer of heat into a closed system at constant volume and temperature: $dQ_{rev}$ = reversible heat transfer.

The classical definition of entropy based on the second law of thermodynamics has given the total differential of entropy in the form of $dQ_{rev}/T$. With a reversible heat transfer into a closed system receiving a differential amount of heat $dQ_{rev}$, the system changes its entropy by the differential amount of $dS$ as shown in Eq. 3.8:

$$dS = \frac{dQ_{rev}}{T} = \frac{dU - dW_{rev}}{T}, \tag{3.8}$$

where $dQ_{rev}$ is the heat reversibly absorbed by the system, $dW_{rev}$ is the work reversibly done to the system, and $dU$ is the change in the internal energy of the system. This classical equation 3.8 is equivalent to the statistical equation 3.7 for the entropy. Figure 3.1 shows the change in entropy due to a reversible transport of heat into a closed system.

In conclusion, entropy is the physical quantity that represents the capacity of distribution of energy over the energy levels of the individual constituent particles in the system. The extensive variable entropy $S$ and the intensive variable the absolute temperature $T$ are *conjugated variables*, whose product $TdS$ represents the heat reversibly transferred into or out of the system. In other words, the reversible transfer of heat into or out of the system is always accompanied by the transfer of entropy.

### 3. 2. Reversible and Irreversible Processes.

A physicochemical change is said to be *reversible*, if it occurs at an infinitesimally small rate without any friction and if both system and surroundings remain in a state of quasi equilibrium: the variables characterizing the system go and return through the same values in the forward and backward changes at an infinitesimally small rate. No change that occurs in nature is reversible, though some real processes can be brought as close as possible to reversible processes. The reversible change is thus regarded as an ideal change which real processes can possibly approach and to which equilibrium thermodynamics can apply. All changes other than the reversible changes are termed *irreversible*; such as changes in volume under a pressure gradient, heat transfer under a temperature gradient, and chemical reactions, all of which take place at a rate of finite magnitude.

In an advancing irreversible process such as a mechanical movement of a body, dissipation of energy for instance from a mechanical form to a thermal form (frictional heat) takes place. The second law of thermodynamics defines the energy dissipation due to irreversible processes in terms of the *creation of entropy* $S_{irr}$ or the creation of *uncompensated heat* $Q_{irr}$.

In a closed system a reversible process creates no entropy so that any change $dS$ in entropy is caused only by an amount $dQ_{rev}$ of heat reversibly transferred from the surroundings as shown in Eqs. 3.8 and 3.9:

$$dS = \frac{dQ_{rev}}{T}, \qquad \text{reversible processes.} \qquad (3.9)$$

An irreversible process, by contrast, creates an amount of entropy so that the total change $dS$ in entropy in a closed system consists not only of an entropy change $dS_{rev}$ due to reversible heat transfer $dQ_{rev}$ from the surroundings but also of an amount of entropy $dS_{irr}$ created by the irreversible process as shown in Eq. 3.10:

$$dS = \frac{dQ_{rev}}{T} + \frac{dQ_{irr}}{T} = \frac{dQ_{rev}}{T} + dS_{irr}, \qquad \frac{dQ_{irr}}{T} = dS_{irr}, \qquad \text{irreversible processes.} \qquad (3.10)$$

This equation 3.10 defines the *creation of uncompensated heat* $Q_{irr}$ and the *creation of entropy* $S_{irr}$:

$$dS_{irr} = \frac{dQ_{irr}}{T} = dS - \frac{dQ_{rev}}{T} > 0, \qquad \text{irreversible processes.} \qquad (3.11)$$

Distinguishing the created entropy $d_eS_{rev}$ from the transferred entropy $d_iS_{irr}$, we express the total change in entropy as the sum of the two parts shown in Fig. 3.2 and Eq. 3.12:

$$dS = d_eS_{rev} + d_iS_{irr}. \qquad (3.12)$$

For a closed system with reversible transfer of heat $dQ_{rev}$ where an irreversible process occurs creating uncompensated heat $Q_{irr}$, these transferred and created parts of entropy are thus given, respectively, in Eq. 3.13:

$$d_eS_{rev} = \frac{dQ_{rev}}{T}, \qquad d_iS_{irr} = \frac{dQ_{irr}}{T} > 0. \qquad (3.13)$$

In an isolated system where no heat transfer occurs into or out of it ($d_eS = 0$), the entropy increases itself whenever the system undergoes irreversible processes: this is one of the expressions of the second law of classical thermodynamics that entropy increases in an isolated system when irreversible processes occur in the system. In a closed system where the transferring entropy can be positive or negative, the total entropy does not necessarily increase with irreversible processes. This is also the case for an open system where the transfer of both heat and substances is allowed to occur into or out of the system. In any type of system, isolated, closed, or open systems, however, the advancement of irreversible processes always causes the creation of entropy in the system.

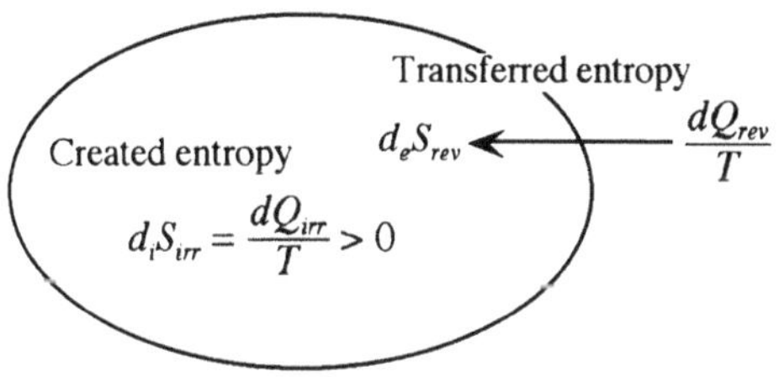


Fig. 3.2. Entropy $d_eS_{rev}$ reversibly transferred from the outside and entropy $d_iS_{irr}$ created by irreversible processes in a closed system.

### 3. 3. The Creation of Entropy and Uncompensated Heat.

As an irreversible process advances in a closed system, the creation of entropy inevitably occurs dissipating a part of the energy of the system in the form of uncompensated heat. The irreversible energy dissipation can be observed, for instance, with the generation of frictional heat in mechanical processes and with the rate-dependent heat generation in chemical reactions different from the reversible heat of reaction. In general, the creation of entropy is always caused by the presence of resistance against the advancement in irreversible processes

We consider a simple chemical reaction, AB $\rightarrow$ A + B, such as $CO_2 \rightarrow CO + 0.5O_2$, in which reacting particles (molecules) distribute their energy among themselves in accord with Boltzmann's distribution law. In order for the reaction to occur, the reacting molecules have to leap over an energy barrier (activation energy) that normally exists along the reaction path from the initial state to the final state of the reaction as shown in Fig. 3.3: this is a flow of reacting molecules through an activated state required for the reaction to proceed.

In the case that the process is reversible in which the initial and the final states are in the same energy level, as shown in Fig. 3.3(a), the energy absorbed by the reacting molecules rising up from the initial state to the activated state equals the energy released when the molecules fall from the activated state down to the final state of the reaction, and hence no net energy dissipation occurs during the reaction.

In the case in which the reaction occurs irreversibly at a finite rate, however, there exists an energy gap between the initial state and the final state of the reaction as shown in Fig. 3.3(b). As the reaction proceeds, then, the amount of energy equivalent to the energy gap dissipates, thereby producing an amount of uncompensated heat and creating an amount of entropy. Usually, the energy diagram of a chemical reaction at constant $T$ and $p$ is expressed in terms of free enthalpy (Gibbs energy) which will be introduced in the following sections. It follows from Fig. 3.3(b) that the energy equivalent to the uncompensated heat created as a result of an irreversible reaction corresponds to the driving force (affinity $A$ = the difference in free enthalpy between the initial and the final states) of the irreversible reaction.

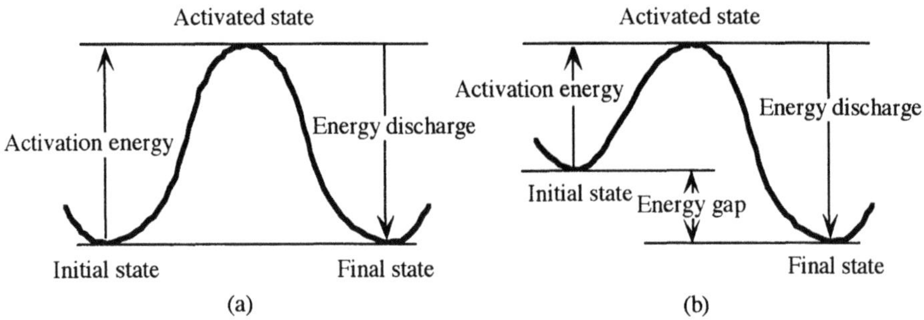


Fig. 3.3. Energy diagrams for (a) reversible process and (b) irreversible process.

According to irreversible thermodynamics [Ref. 2.], the rate of the creation of uncompensated heat, which equals the rate of the creation of entropy times the absolute temperature, is equivalent to the driving force $A$ times the rate $v = d\xi/dt$ of the irreversible reaction as shown in Eq. 3.14 (vid. Eq. 3.39):

$$\frac{dQ_{irr}}{dt} = T\,\frac{dS_{irr}}{dt} = A\,v \geq 0\,. \tag{3.14}$$

In the range in which a linear relationship $v = k\,A$ holds between the driving force $A$ and the rate v of the reaction, Eq. 3.14 yields Eq. 3.15:

$$\frac{dQ_{irr}}{dt} = T\,\frac{dS_{irr}}{dt} = k\,A^2 \geq 0, \tag{3.15}$$

where $1/k$ is a reaction resistance. Equation 3.15 indicates that the rate of the creation of uncompensated heat is proportional to the square of the driving force $A$, the energy gap between the initial state and the final state of the processes. Note that linear reaction kinetics $v = k\,A$ is valid only in the regime close to the reaction equilibrium, beyond which non-linear exponential kinetics usually predominates.

**3. 4. The Creation of Entropy and Thermodynamic Potentials.**

From Eq. 2.3 of the conservation of energy, $dQ_{rev} = dU + pdV$, and Eq. 3.10 of the creation of entropy, $dQ_{rev} = TdS - TdS_{irr}$, we obtain Eq. 3.16 for an infinitesimal advancement of an irreversible process in a *closed* system:

$$dU + p\,dV = T\,dS - T\,dS_{irr}, \tag{3.16}$$

where $dS = dS_{rev} + dS_{irr}$ is the differential of the total entropy $dS$ consisting of the entropy $dS_{rev}$ reversibly transferred and the entropy $dS_{irr}$ irreversibly created in the system. The entropy created by an irreversible process is always positive (plus) $dS_{irr} > 0$. The differential of the internal energy $U$ is then given by Eq. 3.17:

$$dU = T\,dS - p\,dV - T\,dS_{irr}\,. \tag{3.17}$$

We thus see that an irreversible process, if occurring at constant entropy and volume ($dS = 0$ and $dV = 0$), is accompanied by a decrease in the internal energy of the system as shown in Eq. 3.18:

$$T\,dS_{irr} = -\,dU > 0, \quad -\left(U_{final} - U_{initial}\right) > 0; \quad \text{at constant } S \text{ and } V, \tag{3.18}$$

where $\left(U_{final} - U_{initial}\right)$ is the change in internal energy between the initial state and the final state of the irreversible process.

Similarly, for the energy function enthalpy $H = U + pV$ defined in the foregoing chapter we have Eq. 3.19 from Eq. 3.17:

$$d\,H = T\,dS + V\,dp - T\,dS_{irr}\,, \tag{3.19}$$

which yields Eq. 3.20 for an irreversible process at constant entropy $S$ and pressure $p$:

$$T\,dS_{irr} = -\,dH > 0, \quad -\left(H_{final} - H_{initial}\right) > 0; \quad \text{at constant } S \text{ and } p\,. \tag{3.20}$$

This indicates that any irreversible process, if occurring at constant entropy $S$ and pressure $p$, is accompanied by a decrease in the enthalpy from the initial high level $H_{initial}$ toward the final low level $H_{final}$ of the system. From the foregoing we see that the internal energy and enthalpy may play the role of thermodynamic potentials for an irreversible process if occurring under the condition of constant entropy $S$. This condition of constant entropy, however, is unrealistic because entropy $S$ contains both created entropy $S_{irr}$ and transferred entropy $S_{rev}$.

We then introduce two new energy functions called *free energy* $F$ (Helmholtz energy) for the independent variables temperature $T$ and volume $V$, and *free enthalpy* $G$ (Gibbs energy) for the independent variables temperature $T$ and pressure $p$ as defined, respectively, in Eqs. 3.21 and 3.22:

$$F = U - T\,S, \tag{3.21}$$

$$G = U - T\,S + pV = H - T\,S. \tag{3.22}$$

Inserting $F$ and $G$ into Eq. 3.17 yields Eqs. 3.23 and 3.24:

$$d\,F = -\,S\,dT - p\,dV - T\,dS_{irr}, \tag{3.23}$$

$$dG = -\,S\,dT + V\,dp - T\,dS_{irr}. \tag{3.24}$$

For an irreversible process in an isothermal system at constant $V$, we then obtain Eq. 3.25:

$$T\,dS_{irr} = -\,dF > 0, \quad -\left(F_{final} - F_{initial}\right) > 0, \tag{3.25}$$

while in an isothermal system at constant pressure $p$, Eq. 3.26 holds:

$$T\,dS_{irr} = -\,dG > 0, \quad -\left(G_{final} - G_{initial}\right) > 0. \tag{3.26}$$

Obviously, the energy functions of free energy $F$ and free enthalpy $G$ play the role of thermodynamic potentials for an irreversible process to occur in isothermal systems at constant volume and constant pressure, respectively. In general, the energy functions of $F$, and $G$ can be used as the thermodynamic potentials to indicate the direction of an irreversible processes to occur under the condition that their respective characteristic variables remain constant.

As mentioned above, free energy $F$ is occasionally called the Helmholtz energy, and free enthalpy $G$ is frequently called the Gibbs energy. These two energy functions $F$ and $G$ correspond to the amounts of energy that are freed from the restriction of entropy and hence can be fully utilized for irreversible processes to occur at constant temperature.

### 3. 5. Affinity of Irreversible Processes.

We now consider a simple system in which equilibrium is already established with respect to temperature and pressure and in which, on the other hand, equilibrium is not attained with respect to the redistribution of substances susceptible to chemical reactions, nor with respect to any changes being characterized by the parameter $\xi$, the extent of reaction shown in Eq. 1.11. Let us first consider a system in which a single chemical reaction takes place in an irreversible way. Suppose that in an infinitesimal time interval the value of $\xi$ changes by an amount $d\xi$, producing then an amount of uncompensated heat $dQ_{irr}$ and hence an amount of created entropy $dS_{irr}$. We now introduce a new energy function called the *affinity* $A$ of an irreversible process defined by the relation shown in Eq. 3.27. Namely, the differential of the irreversibly dissipated energy (uncompensated heat) $dQ_{irr}$ equals the affinity $A$ times the differential of the extent of reaction $d\xi$:

$$dQ_{irr} = T\,dS_{irr} = A\,d\xi > 0. \tag{3.27}$$

Equation 3.27 is called *De Donder's inequality*.

The affinity is expressed as a function of independent variables such as $A(T,V,\xi)$ and $A(T,p,\xi)$. For the characteristic variables $T$, $V$, and $\xi$, we obtain from Eqs. 2.5 (where $Q = Q_{rev}$), 3.11 and 3.27 the following equation 3.28 for the affinity $A$ of the reaction:

$$-A = \left(\frac{\partial U}{\partial \xi}\right)_{T,V} - T\left(\frac{\partial S}{\partial \xi}\right)_{T,V}, \tag{3.28}$$

while for the characteristic variables of $T$, $p$, and $\xi$, we further obtain from Eqs. 2.14 (where $Q = Q_{rev}$), 3.11 and 3.27 the following Eq. 3.29 for the affinity $A$ of the reaction at constant temperature and pressure:

$$-A = \left(\frac{\partial H}{\partial \xi}\right)_{T,p} - T\left(\frac{\partial S}{\partial \xi}\right)_{T,p}. \tag{3.29}$$

These two equations show that the affinity depends not only on the internal energy $U$ or the enthalpy $H$ but also on the entropy $S$.

Combining Eq. 3.27 with Eqs. 3.17, 3.19, 3.23, and 3.24, we obtain the following four equations:

$$dU = T\,dS - p\,dV - A\,d\xi, \tag{3.30}$$

$$dH = T\,dS + V\,dp - A\,d\xi, \tag{3.31}$$

$$dF = -S\,dT - p\,dV - A\,d\xi, \tag{3.32}$$

$$dG = -S\,dT + V\,dp - A\,d\xi. \tag{3.33}$$

These four thermodynamic energy functions of state $U = U(S,V,\xi)$, $H = H(S,p,\xi)$, $F = F(T,V,\xi)$, and $G = G(T,p,\xi)$ are called the *thermodynamic potentials* for the characteristic variables $S$ and $V$; $S$ and $p$; $T$ and $V$; and $T$ and $p$; respectively.

An irreversible process advances, if its affinity is positive ($A > 0$), and it finally reaches the equilibrium state where the affinity becomes zero ($A = 0$). This indicates that the advancement in an irreversible process is accompanied by decreasing thermodynamic potentials. As shown in Fig. 3.4, an irreversible process proceeds in the direction in which the thermodynamic potentials of the process decrease. In principle, the affinity decreases as the irreversible process proceeds.

The affinity of irreversible processes, as mentioned above, is related to the thermodynamic potentials $U$, $H$, $F$, and $G$ under the conditions that their respective characteristic variables are kept constant. From Eqs. 3.30, 3.31, 3.32, and 3.33, we obtain the partial differentials of these thermodynamic potentials with respect to their respective characteristic variables as shown in Eqs. 3.34, 3.35, 3.36, and 3.37:

$$\left(\frac{\partial U}{\partial S}\right)_{V,\xi}=T\,,\quad \left(\frac{\partial U}{\partial V}\right)_{S,\xi}=-p\,,\quad \left(\frac{\partial U}{\partial \xi}\right)_{S,V}=-A, \tag{3.34}$$

$$\left(\frac{\partial H}{\partial S}\right)_{p,\xi}=T\,,\quad \left(\frac{\partial H}{\partial p}\right)_{S,\xi}=-V\,,\quad \left(\frac{\partial H}{\partial \xi}\right)_{S,p}=-A, \tag{3.35}$$

$$\left(\frac{\partial F}{\partial T}\right)_{V,\xi}=-S\,,\quad \left(\frac{\partial F}{\partial V}\right)_{T,\xi}=-p\,,\quad \left(\frac{\partial F}{\partial \xi}\right)_{T,V}=-A, \tag{3.36}$$

$$\left(\frac{\partial G}{\partial T}\right)_{p,\xi}=-S\,,\quad \left(\frac{\partial G}{\partial p}\right)_{T,\xi}=V\,,\quad \left(\frac{\partial G}{\partial \xi}\right)_{T,p}=-A. \tag{3.37}$$

In these equations we see the regularity that the partial differential of these four thermodynamic potentials with respect to their respective extensive variables gives us their conjugated intensive variables and vice versa. We thus obtain minus the affinity of an irreversible process in terms of the partial differentials of $U$, $H$, $F$, and $G$ with respect to the extent of reaction; affinity is an extensive variable.

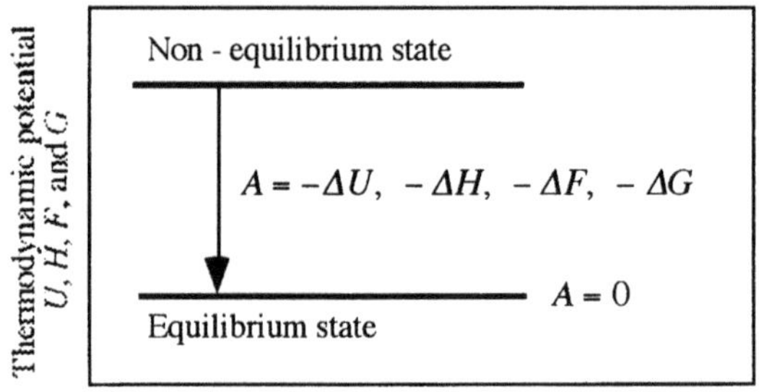


Fig. 3.4. An irreversible process occurs reducing its affinity from a state of high thermodynamic potentials to an equilibrium state of low thermodynamic potentials.

The differentials of the energy functions are complete differentials with the property that the mixed second order differentials are equal to each other. This leads to important relations as exemplified for the free enthalpy by Eq. 3.38 as obtained from Eq. 3.37:

$$\left(\frac{\partial A}{\partial T}\right)_{p,\xi}=\left(\frac{\partial S}{\partial \xi}\right)_{T,p},\quad \left(\frac{\partial A}{\partial p}\right)_{T,\xi}=-\left(\frac{\partial V}{\partial \xi}\right)_{T,p},\quad \left(\frac{\partial S}{\partial p}\right)_{T,\xi}=-\left(\frac{\partial V}{\partial T}\right)_{p,\xi}. \tag{3.38}$$

From Eq. 3.27 we have for an irreversible process the rate of energy dissipation $dQ_{irr}/dt = T\,dS_{irr}/dt$ equal to the affinity $A$ times the rate $d\xi/dt = \mathrm{v}$ as shown in Eq. 3.39:

$$\frac{dQ_{irr}}{dt} = \frac{T\,dS_{irr}}{dt} = A\,\frac{d\xi}{dt} = A\,v > 0. \tag{3.39}$$

We thus see that the affinity always has the same sign as the rate of the process. If the affinity is positive $A > 0$, the rate must be positive $v > 0$ indicating that the irreversible process proceeds in the forward direction; whereas, if the affinity is negative $A < 0$, the rate must be negative $v < 0$ meaning that the process proceeds in the backward direction. When the affinity decreases to zero $A = 0$, the rate of process also decreases to zero and the process is in equilibrium. This property of affinity is characteristic of all kinds of irreversible processes such as the transfer of heat under a gradient of temperature and chemical reactions under a gradient of thermodynamic potentials.

Equation 3.39 holds valid for the system in which only a single process or reaction is occurring. In a system in which multiple chemical reactions are simultaneously occurring, Eq. 3.27 for the uncompensated heat can be expressed by the sum of the products of all independent affinities and their conjugated reaction rates as given in Eq. 3.40:

$$\frac{dQ_{irr}}{dt} = \frac{T\,dS_{irr}}{dt} = \sum_i A_i\,\frac{d\xi_i}{dt} = \sum_i A_i\,v_i > 0, \tag{3.40}$$

where $A_i$ is the affinity of the $i$th reaction and $\xi_i$ is the corresponding extent of reaction.

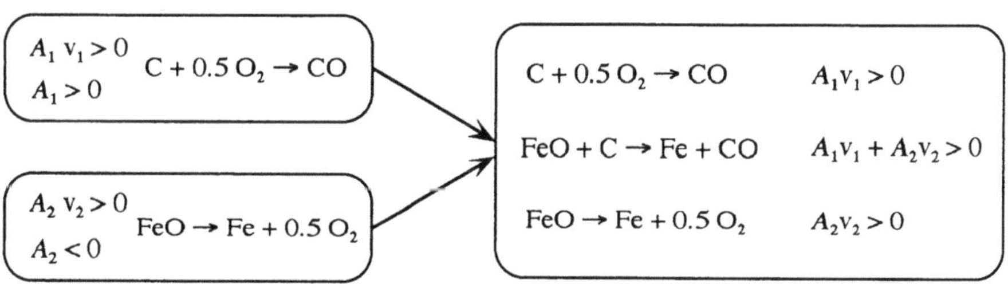


Fig. 3.5. Energy transfer from a coupling reaction $C + 0.5\,O_2 \rightarrow CO$ to a coupled reaction $FeO \rightarrow Fe + 0.5\,O_2$ for a combined reaction $FeO + C \rightarrow Fe + CO$.

We also notice that the creation of entropy in various reactions occurring simultaneously is positive as a whole, though it may be positive or negative for individual reactions. Thus, in a system in which two chemical reactions occur, it is possible that $A_1 v_1 > 0$ for one reaction and $A_2 v_2 < 0$ for the other, provided that $A_1 v_1 + A_2 v_2 > 0$. In such a case we call reaction 1 the *coupling* reaction which proceeds producing an amount of created entropy (uncompensated heat) under its positive affinity, and reaction 2 is the *coupled* reaction which proceeds absorbing the created entropy (uncompensated heat) under its negative affinity. We thus see that the transfer of energy from the coupling reaction 1 to the coupled reaction 2 makes it possible for the latter to proceed even though its affinity is negative. The rate of a coupled reaction, however, must be within the limit that $v_2 < (A_1 v_1 / A_2)$ shown in Eq. 3.40. As an example of the thermodynamic coupling of chemical reactions, we have the combination of carbon oxidation and iron oxide reduction as shown in Fig. 3.5. Such coupling-and-coupled reactions will again be discussed in terms of *exergy* in chapter 11.

### 3. 6. Entropy of Pure Substances.

The entropy of a pure substance is a function of temperature $T$ and pressure $p$. Equations 2.13 and 3.9 yield the total differential of the molar entropy $ds$ of a pure substance:

$$ds = \frac{dh - v\,dp}{T} = \frac{c_p}{T}\,dT - \frac{v}{T}\,dp, \tag{3.41}$$

where $c_p$, $h$, and $v$ are the molar heat capacity at constant pressure, the molar enthalpy, and the molar volume of the pure substance, respectively. We then obtain Eq. 3.42 for the entropy of a pure substance:

$$s(T,p) = s(0,0) + \int_0^T \frac{c_p(T,0)}{T}\,dT - \int_0^p \frac{v}{T}\,dp, \tag{3.42}$$

where $s(0,0)$ is the molar entropy of the pure substance extrapolated to $p = 0$ and $T = 0$. The third law of thermodynamics, called the *Nernst heat theorem*, assumes that the entropy of the condensed phase of a perfect crystal may be equated with zero at the zero absolute temperature, $s(0,0) \rightarrow 0$: No energy fluctuation occurs at $T = 0$ giving $\Omega(N,V,U) = 1$ in Eq. 3.6 and hence entropy is zero.

On the right hand side of Eq. 3.42, the second term is the thermal part and the third term is the pressure-dependent part of the molar entropy. The entropy of a pure substance thus consists of the thermal part and the pressure-dependent part. Under ordinary conditions, however, the latter is so small compared with the former that we may regard the entropy as independent of pressure for condensed substances particularly (vid. Eqs. 7.29 and 7.30). For gaseous substances a slight change in entropy results from a change in pressure, $s(T,p) = s^{*}(T,p^0) - R\ln(p/p^0)$ where $p^0$ is a reference pressure, as will be shown in section 3.8.

From Eq. 3.42 we obtain the molar entropy of a pure substance in the gas state at constant pressure as shown in Eq. 3.43:

$$s^g = \int_0^{T_f} \frac{c_p^s}{T}dT + \frac{\Delta_f h}{T_f} + \int_{T_f}^{T_v} \frac{c_p^l}{T}dT + \frac{\Delta_v h}{T_v} + \int_{T_v}^{T} \frac{c_p^g}{T}dT - R\ln\left(\frac{p}{p^0}\right), \tag{3.43}$$

where $c_p^g$, $c_p^l$, and $c_p^s$ are the molar heat capacities of the substance at constant pressure in the gas, liquid, and solid states, respectively; $\Delta_f h$ and $\Delta_v h$ are the heat of fusion and the heat of evaporation, respectively; and $T_f$ and $T_v$ are the melting temperature and the boiling temperature, respectively. In Eq. 3.43 the first term on the right hand side is the thermal part of the molar entropy for the solid state, the second term is for the melting, the third term is for the liquid state, the fourth term is for the boiling, the fifth term is for the thermal part of the gas state, and the last term is the pressure-dependent part of the molar entropy in the ideal gas state.

Figure 3.6 shows schematically the molar entropy of a pure substance as a function of temperature. If a structural transformation occurs in the solid state, an additional increase in the molar entropy comes from the heat of the transformations. As shown in the figure, the molar entropy of a pure substance increases with increasing temperature. In chemical handbooks we see the tabulated numerical values of the molar entropy calculated for a number of pure substances in the standard state at temperature 298 K and pressure 101.3 kPa. A few of them will be listed as the standard molar entropy, $s_i^0$, in Table 5.1. Note that the molar entropy thus calculated based on the third law of thermodynamics is occasionally called "*absolute*" entropy.

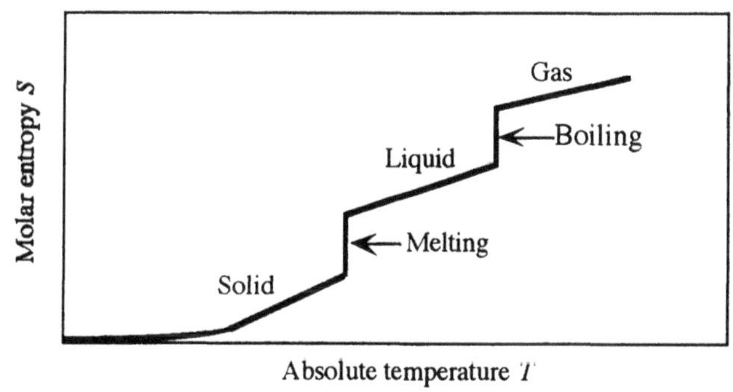


Fig. 3.6. Molar entropy of a pure substance as a function of temperature.

### 3. 7. Entropy of Heat Transfer.

Let us now consider a steady flow of heat $dQ(irr)$ that occurs irreversibly between a phase at a high temperature $T_1$ and a phase at a low temperature $T_2$ in a closed system as shown in Fig. 3.7. The phase 1 continuously receives heat $dQ = T_1 dS_1$ in a reversible way from the surroundings at temperature $T_1$ and the phase 2 continuously releases heat $dQ = T_2 dS_2$ into the surroundings at temperature $T_2$. In the steady state no change occurs in the state property of the system except an increase in entropy $dS_{irr}$ due to the irreversible heat transfer $dQ(irr) = dQ$:

$$dS_{irr} = dS_2 - dS_1 = \left(\frac{1}{T_2} - \frac{1}{T_1}\right) dQ(irr) > 0, \qquad dQ(irr) = \frac{T_1 T_2}{T_1 - T_2}\, dS_{irr}, \tag{3.44}$$

where $dS_1$ and $dS_2$ are the entropy at $T_l$ and the entropy at $T_2$ due to the steady flow of heat $dQ\ (= dQ(irr))$ in a reversible way between the system and the surroundings, respectively. Equation 3.44 shows the relation between the amount of heat $dQ(irr)$ irreversibly transferred and the amount of entropy $dS_{irr}$ irreversibly created in the steady flow of heat between the two different temperatures. In this case the irreversibly created entropy is continuously released from the low temperature phase to the surroundings so that no accumulation of created entropy occurs in the system.

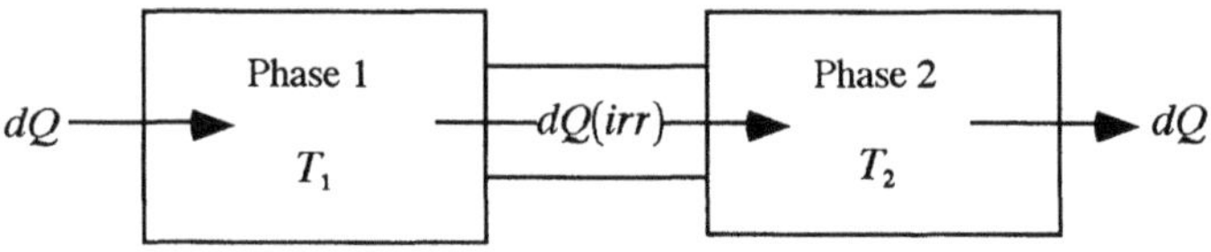


Fig. 3.7. Irreversible steady flow of heat from phase 1 at high temperature $T_1$ to phase 2 at low temperature $T_2$ in a closed system.

Heat transfer between two different temperatures can be carried out in a reversible way by using a reversible heat engine or heat pump. In this case, however, a part of the transferring heat converts into work or a part of the transferring heat is created by work. The *heat engine* is a closed system of a gas, in which a quantity of heat $dQ_1$ is absorbed from an outside heat source at a high temperature $T_1$ and preforms a quantity of work $dW$ to the exterior of the system releasing a quantity of heat $dQ_2$ less than $dQ_1$ into an outside heat reservoir at a low temperature $T_2$. On the other hand, a closed system of a gas is called a *heat pump* or an *inverse heat engine*, when it receives a quantity of work $dW$ from the outside and takes up a quantity of heat $dQ_2$ from an outside heat reservoir at a low temperature $T_2$ bringing a quantity of heat $dQ_1$ more than $dQ_2$ into an outside heat reservoir at a high temperature $T_1$. Figure 3.8 shows the processes that occur in a heat engine and a heat pump. One of the ideal heat engines operating in a reversible way is known as *Carnot's heat engine*, in which two adiabatic and two isothermal processes constitute what is called the Carnot cycle.

From the first law (energy conservation) of thermodynamics we have $dQ_1 = dW + dQ_2$, and the second law (entropy creation) of thermodynamics gives us $(dQ_1/T_1) + (dQ_2/T_2) \geq 0$, where equality is for a reversible heat engine and inequality for an irreversible one. We then have the efficiency $\lambda_{1\rightarrow 2} = (dW/dQ_1)$ for the reversible heat engine and the efficiency $\lambda_{2\rightarrow 1} = (dQ_1/dW)$ for the reversible heat pump as shown, respectively, in Eq. 3.45:

$$\lambda_{1\rightarrow 2} = \frac{|dW|}{dQ_1} = \frac{T_1 - T_2}{T_1}, \qquad \lambda_{2\rightarrow 1} = \frac{|dQ_1|}{dW} = \frac{T_1}{T_1 - T_2}. \tag{3.45}$$

No creation of entropy and uncompensated heat occurs in the reversible heat engine and pomp, and hence Eq. 3.45 gives the maximum efficiency theoretically attainable for heat engines and heat pumps. This equation also shows that thermal energy (heat) can not be wholly converted into work and that its conversion efficiency depends on the temperature at which the thermal energy is reserved.

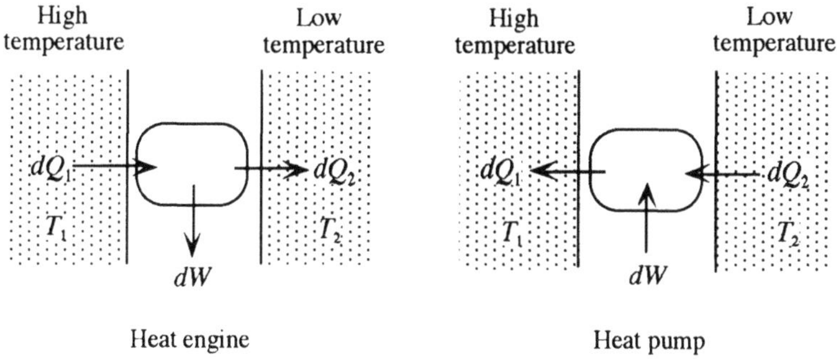


Fig. 3.8. Processes occurring in a heat engine and in a heat pump.

This type of reversible heat transfer by means of a heat engine has its affinity $A$, which is equivalent to the maximum work $W$ obtainable with the engine as expressed by Eq. 3.46:

$$A = W = \frac{T_1 - T_2}{T_1} Q_1 . \qquad (3.46)$$

This is the maximum available energy that we can obtain from an amount of heat $Q_1$ at a temperature $T_1$.

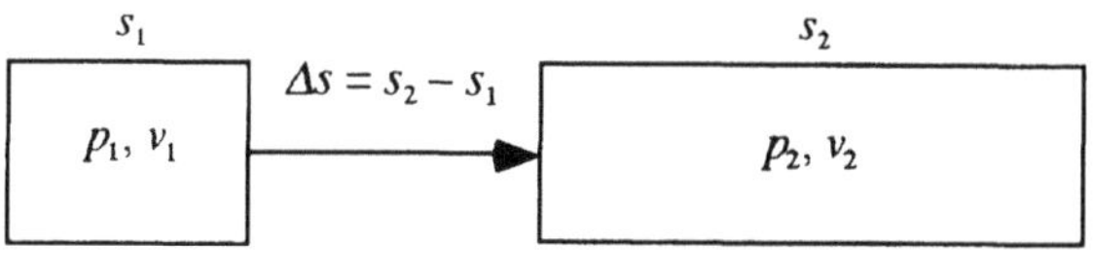


Fig. 3.9. Isothermal expansion of one mole of an ideal gas resulting in an entropy increase.

### 3. 8. Entropy of Gas Expansion.

Let us now discuss the entropy of gas expansion in a closed system. Equation 3.42 gives us the molar entropy of an ideal gas at constant temperature $T$ as shown in Eq. 3.47:

$$s(T, p) = s^{*}(T, p^0) - R \ln \frac{p}{p^0}, \quad s(T, v) = s^{**}(T, v^0) + R \ln \frac{v}{v^0}, \quad s^{*}(T) = s^{**}(T) + R \ln RT, \qquad (3.47)$$

where $v$ is the molar volume of the gas, $p$ is the pressure of the gas, and $p^0$ and $v^0$ are a reference pressure and a reference molar volume, respectively.

If the gas expansion takes place isothermally as shown in Fig. 3.9, the molar entropy of the gas then increases with increasing volume and decreases with increasing pressure as shown in Eq. 3.48:

$$\Delta s = R \ln \frac{v_2}{v_1} = -R \ln \frac{p_2}{p_1}, \tag{3.48}$$

where $\Delta s = s_2 - s_1$ is the change in the molar entropy of a gas caused by the gas expansion from an initial state $(p_1, v_1)$ to a final state $(p_2, v_2)$ at constant temperature.

## 3. 9. Entropy of Mixing.

The mixing of substances is an irreversible process that takes place creating entropy in the system. The entropy thus created is defined as the *entropy of mixing* $S^M$. Suppose two different ideal gases with different volumes $V_1$ and $V_2$ are mixed isothermally at a constant pressure $p$ to make a single mixture system with a volume $V_1 + V_2$ as shown in Fig. 3. 10. The overall entropy $S^I$ of both individual systems before the mixing is obtained from Eq. 3.47 as shown in Eq. 3.49:

$$S^{\mathrm{I}}(T, p) = \sum_i n_i \left\{ s_i^*(T, p_0) - R \ln \frac{p}{p_0} \right\}. \tag{3.49}$$

On mixing the gases we obtain the entropy $S^{\mathrm{F}}$ of the mixed gases as expressed by Eq. 3.50 as a function of the partial pressure $p_i$ $(p = \Sigma_i p_i)$ and the molar fraction $x_i = p_i / p$ of the constituent gases:

$$S^{\mathrm{F}}(T, p) = \sum_i n_i \left\{ s_i^*(T, p_0) - R \ln \frac{p_i}{p_0} \right\} = \sum_i n_i \left\{ s_i^*(T, p_0) - R \ln \frac{p}{p_0} - R \ln x_i \right\}$$

$$= S^{\mathrm{I}}(T, p) - \sum_i n_i R \ln x_i. \tag{3.50}$$

The entropy of mixing $S^M$ is thus given in general in the form of Eq. 3.51:

$$S^M = S^{\mathrm{F}}(T, p) - S^{\mathrm{I}}(T, p) = -\sum_i n_i R \ln x_i > 0. \tag{3.51}$$

The mixing of gases at constant pressure may also be regarded for each constituent gas as an expansion of its volume decreasing its partial pressure $(p \rightarrow p_i)$, and hence the entropy of mixing can also be obtained from Eq. 3.48 for the mixing of two gases as shown in Eq. 3.52:

$$\Delta S = S^M = -n_1 R \ln \frac{p_1}{p} - n_2 R \ln \frac{p_2}{p} = -n_1 R \ln x_1 - n_2 R \ln x_1 > 0. \tag{3.52}$$

The entropy of mixing is generated not only in the gas state (gas mixtures) but also in the states of liquids (liquid solutions) and solids (solid solutions).

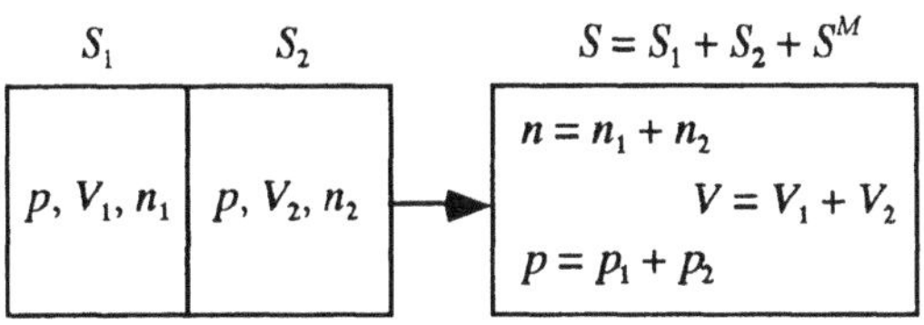


$$S^M = -n_1 R \ln x_1 - n_2 R \ln x_2 > 0$$

Fig. 3.10. Entropy of mixing of two gases to form a gas mixture at constant pressure and temperature.

# CHAPTER 4

# AFFINITY IN IRREVERSIBLE PROCESSES

The affinity of irreversible processes is a thermodynamic function of state related to the creation of entropy and uncompensated heat during the processes. The second law of thermodynamics indicates that all irreversible processes advance in the direction of creating entropy and decreasing affinity. This chapter examines the property affinity in chemical reactions and the relation between the affinity and various other thermodynamic quantities.

## 4. 1. Affinity in Chemical Reactions.

The concept of affinity introduced in the foregoing chapter (section 3.5) can apply to all the physicochemical changes that occur irreversibly. Let us now discuss the physical meaning of the affinity of chemical reactions. As mentioned in the foregoing, we have in Eq. 3.27 the fundamental inequality in entropy balance of irreversible processes as shown in Eq. 4.1:

$$dQ_{irr} = T\,dS_{irr} = A\,d\xi \geq 0, \qquad A\,\frac{d\xi}{dt} = A\,v \geq 0. \tag{4.1}$$

The inequality in this equation is for irreversible reactions that occur spontaneously, while the equality is for reversible reactions in quasi-equilibrium. The inequality equation 4.1 is in fact the most important property of the affinity showing that the affinity always has the same sign as that of the rate of reaction at any instance during the reaction.

In Eqs. 3.30 to 3.33, we have seen a series of equations for the various thermodynamic potentials as functions of the affinity as follows:

$$dU = T\,dS - p\,dV - A\,d\xi, \tag{4.2}$$

$$dH = T\,dS + V\,dp - A\,d\xi, \tag{4.3}$$

$$dF = -S\,dT - p\,dV - A\,d\xi, \tag{4.4}$$

$$dG = -S\,dT + V\,dp - A\,d\xi. \tag{4.5}$$

These equations give us the affinity as the partial differential of the thermodynamic potentials with respect to the extent of reaction as shown in Eq. 4.6:

$$A = -\left(\frac{\partial U}{\partial \xi}\right)_{S,V} = -\left(\frac{\partial H}{\partial \xi}\right)_{S,p} = -\left(\frac{\partial F}{\partial \xi}\right)_{T,V} = -\left(\frac{\partial G}{\partial \xi}\right)_{T,p}. \tag{4.6}$$

This equation indicates that the affinity corresponds to the thermodynamic potentials of $U$ and $H$ under the conditions of constant entropy $S$ and to the thermodynamic potentials of $F$ and $G$ under the conditions of constant temperature $T$.

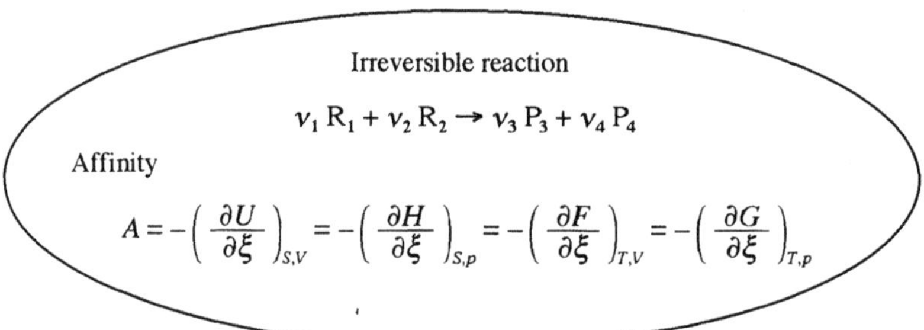


Fig. 4.1. Affinity in a chemical reaction.

If the affinity is zero $A = 0$, no irreversible reaction advances and the system is in equilibrium. Then the equations from 4.2 to 4.5, if excluding the third terms on their right hand side, represent the fundamental properties of thermodynamic potentials $U$, $H$, $F$, and $G$ in the state of reaction equilibrium, i.e. the state in which no physicochemical change occurs.

**4. 2. Affinity and Heat of Reaction.**

Equations 3.28 and 3.29 have shown the relationship between the affinity $A$ and the heats of reaction $\partial U/\partial \xi$ at constant volume and $\partial H/\partial \xi$ at constant pressure as shown in Eq. 4.7:

$$-A = \left(\frac{\partial U}{\partial \xi}\right)_{T,V} - T\left(\frac{\partial S}{\partial \xi}\right)_{T,V}, \qquad -A = \left(\frac{\partial H}{\partial \xi}\right)_{T,p} - T\left(\frac{\partial S}{\partial \xi}\right)_{T,p}. \tag{4.7}$$

In the case in which the second entropy term on the right hand side of the above equations is significantly small compared with the first energy or enthalpy term (i.e. the system is at very low temperature), the affinity is nearly equal to the heat of reaction: $A \approx -(\partial U/\partial \xi)_{T,V}$ or $A \approx -(\partial H/\partial \xi)_{T,p}$. If an irreversible reaction ($A > 0$) has the enthalpy term larger than the entropy term, $(\partial H/\partial \xi)_{T,p} > T(\partial S/\partial \xi)_{T,p}$, the enthalpy term is negative $(\partial H/\partial \xi)_{T,p} < 0$ and hence the reaction is exothermic.

Reminding that for the independent state variables $p$, $T$, and $\xi$ the following relations hold:

$$\left(\frac{\partial H}{\partial T}\right)_{p,\xi} = C_{p,\xi}, \quad \left(\frac{\partial S}{\partial T}\right)_{p,\xi} = \frac{C_{p,\xi}}{T}, \quad \left[\frac{\partial}{\partial T}\left(\frac{\partial H}{\partial \xi}\right)_{p,T}\right]_{p,\xi} = \left(\frac{\partial C_{p,\xi}}{\partial \xi}\right)_{p,T},$$

$$\left[\frac{\partial}{\partial T}\left(\frac{\partial S}{\partial \xi}\right)_{p,T}\right]_{p,\xi} = \left[\frac{\partial}{\partial \xi}\left(\frac{\partial S}{\partial T}\right)_{p,\xi}\right]_{p,T} = \frac{1}{T}\left(\frac{\partial C_{p,\xi}}{\partial \xi}\right)_{p,T},$$

and that the similar equations also hold for the independent variables $V$, $T$, and $\xi$, we obtain Eq. 4.8 from Eq. 4.7 [Ref. 1.]:

$$\left[\frac{\partial}{\partial T}\left(\frac{A}{T}\right)\right]_{V,\xi} = \frac{1}{T^2}\left(\frac{\partial U}{\partial \xi}\right)_{T,V}, \qquad \left[\frac{\partial}{\partial T}\left(\frac{A}{T}\right)\right]_{p,\xi} = \frac{1}{T^2}\left(\frac{\partial H}{\partial \xi}\right)_{T,p}. \tag{4.8}$$

Integrating the second equation of Eq. 4.8 yields Eq. 4.9:

$$\left[\frac{A(T,p,\xi)}{T} - \frac{A(T_0,p,\xi)}{T_0}\right]_{p,\xi} = \int_{T_0}^{T} \frac{1}{T^2}\left(\frac{\partial H}{\partial \xi}\right)_{T,p} dT. \tag{4.9}$$

We can make use of this equation 4.9 to estimate the affinity $A(T,p,\xi)$ as a function of temperature $T$ from a known affinity value $A(T_0,p,\xi)$ at a specified temperature $T_0$, provided that we know the molar heat capacities of the reactants and products.

From Eqs. 3.38 and 4.8 we obtain the total differential of $A(T,p,\xi)/T$ shown in Eq. 4.10:

$$d\left(\frac{A}{T}\right) = \frac{1}{T^2}\left(\frac{\partial H}{\partial \xi}\right)_{T,p} dT - \frac{1}{T}\left(\frac{\partial V}{\partial \xi}\right)_{T,p} dp + \frac{1}{T}\left(\frac{\partial A}{\partial \xi}\right)_{T,p} d\xi. \tag{4.10}$$

Using the three symbols $(\partial A/\partial \xi)_{T,p} = a_{T,p}$, $(\partial V/\partial \xi)_{T,p} = v_{T,p}$, and $(\partial H/\partial \xi)_{T,p} = h_{T,p}$, we may put Eq. 4.10 into the alternative form shown in Eq. 4.11:

$$dA = \frac{A + h_{T,p}}{T} dT - v_{T,p}\, dp + a_{T,p}\, d\xi. \tag{4.11}$$

Equations 4.10 and 4.11 hold valid for a single reaction. For multiple reactions occurring simultaneously in the system the third term on the right hand side of Eq. 4.11 consists not only of the sum of individual reactions but also of the sum of the interactions among the reactions. Since the interaction between reaction $i$ and reaction $j$ is given in Eq. 4.12:

$$\left(\frac{\partial A_i}{\partial \xi_j}\right)_{T,p} = -\left(\frac{\partial^2 G}{\partial \xi_i\,\partial \xi_j}\right)_{T,p} = \left(\frac{\partial A_j}{\partial \xi_i}\right)_{T,p}, \tag{4.12}$$

we have in place of Eq. 4.10 Eq. 4.13 for simultaneous multiple reactions:

$$d\left(\frac{A_i}{T}\right) = \frac{1}{T^2}\left(\frac{\partial H}{\partial \xi_i}\right)_{T,p} dT - \frac{1}{T}\left(\frac{\partial V}{\partial \xi_i}\right)_{T,p} dp + \frac{1}{T}\sum_j\left(\frac{\partial A_i}{\partial \xi_j}\right)_{T,p} d\xi_j. \tag{4.13}$$

This equation 4.13 can be used for studying the stability of chemical equilibria of multiple chemical reactions.

## 4. 3. The Average Heat of Reaction.

The heat $Q(\xi)$ absorbed in or released out of a closed system in which a chemical reaction occurs is in principle a function of the extent of reaction $\xi$, temperature $T$, and pressure $p$ (or volume $V$) of the reaction system. If we fix the condition at which the reaction occurs, all the variables defining the state of the reaction system will be definite functions of $\xi$. We now introduce a symbol $q(\xi)$ denoting the heat received by the system when the reaction proceeds by an extent $d\xi$ as defined in Eq. 4.14:

$$q(\xi) = \frac{dQ(\xi)}{d\xi}. \tag{4.14}$$

The function of $q(\xi)$ is the heat of reaction relative to the chemical reaction under consideration [Ref. 1.]. We further define the average heat of reaction for the change from $\xi_0$ to $\xi_1$ as expressed by Eq. 4.15. Integrating Eq. 4.14 for one equivalent extent of the reaction, we obtain the average heat of reaction $Q^{\text{mean}}$ for the chemical change as shown in Eq. 4.15:

$$Q^{mean} = \frac{1}{\xi_1 - \xi_0}\int_0^1 q(\xi)\, d\xi. \tag{4.15}$$

We usually call $Q^{\text{mean}}$ the *average heat of reaction* when the chemical change occurs as much as one equivalent extent of reaction, $\xi_1 - \xi_0 = 1$.

For a system at constant temperature and volume, as described in Eqs. 2.5. 2.6 and 2.7, the differential $(\partial U/\partial \xi)_{T,V} d\xi = u_{T,V} d\xi$ gives us the average heat of reaction at constant $T$ and $V$ as shown in Eq. 4.16:

$$Q^{\text{mean}}_{T,V} = \int_0^1 \left(\frac{\partial U}{\partial \xi}\right)_{T,V} d\xi = \int_0^1 u_{T,V}\, d\xi = \left(U_{\xi=1} - U_{\xi=0}\right)_{T,V} = (\Delta U)_{T,V}. \tag{4.16}$$

This equation indicates that the average heat of reaction $Q^{\text{mean}}_{T,V}$ at constant $T$ and $V$ equals the change in the internal energy $(\Delta U)_{T,V}$ for one equivalent extent of the reaction.

Similarly, from Eqs. 2.13 and 2.19 we obtain Eq. 4.17 for the average heat of reaction at constant $T$ and $p$:

$$Q_{T,p}^{\text{mean}} = \int_0^1 \left(\frac{\partial H}{\partial \xi}\right)_{T,p} d\xi = \int_0^1 h_{T,p} d\xi = (H_{\xi=1} - H_{\xi=0})_{T,p} = (\Delta H)_{T,p}, \tag{4.17}$$

indicating that the average heat of reaction $Q_{T,p}^{\text{mean}}$ at constant $T$ and $p$ is equivalent to the decrease in the enthalpy for one equivalent extent of the reaction.

The reaction is exothermic (heat-releasing) when the heat of reaction is negative, while it is endothermic (heat-absorbing) when the heat of reaction is positive.

## 4. 4. The Average Affinity of Reaction

The affinity of a reaction, $A(\xi)$, is also a function of the extent of reaction $\xi$. On integrating the reaction affinity $A(\xi)$ for one equivalent extent of the reaction from $\xi_0 = 0$ to $\xi_1 = 1$, we define the *average affinity of reaction*, $A^{\text{mean}}$, as shown in Eq. 4.18:

$$A^{mean} = \frac{1}{\xi_1 - \xi_0} \int_0^1 A(\xi)\, d\xi . \tag{4.18}$$

For a chemical reaction at constant temperature and volume, Eqs. 3.21, 4.7 and 4.18 give the average affinity of the reaction as shown in Eq. 4.19:

$$A_{T,V}^{\text{mean}} = -\int_0^1 \left(\frac{\partial F}{\partial \xi}\right)_{T,V} d\xi = -(F_{\xi=1} - F_{\xi=0})_{T,V} = -(\Delta F)_{T,V}, \tag{4.19}$$

indicating that the average affinity $A_{T,V}^{\text{mean}}$ at constant temperature $T$ and volume $V$ is equal to the decrease $-(\Delta F)_{T,V}$ in the free energy (Helmholtz energy) that occurs during the advancement of one equivalent extent of the reaction.

Similarly, for a chemical reaction at constant temperature $T$ and pressure $p$ Eqs. 3.22, 4.7 and 4.18 give the average affinity of the reaction as shown in Eq. 4.20:

$$A_{T,p}^{\text{mean}} = -\int_0^1 \left(\frac{\partial G}{\partial \xi}\right)_{T,p} d\xi = -(G_{\xi=1} - G_{\xi=0})_{T,p} = -(\Delta G)_{T,p}. \tag{4.20}$$

Equation 4.20 indicates that the average affinity $A_{T,p}^{\text{mean}}$ at constant $T$ and $p$ is equivalent to the decrease $-(\Delta G)_{T,p}$ in the free enthalpy (Gibbs energy) that occurs during the advancement of one equivalent extent of the reaction.

The physical quantity that we usually call the affinity of a reaction corresponds to the average affinity of the reaction. Generally, the affinity of a reaction at constant $T$ and $V$ differs numerically from that at constant $T$ and $p$, as compared to the heat of reaction whose numerical value depends on whether the reaction occurs at constant $V$ or at constant $p$.

Integrating Eq. 4.7 from $\xi = 0$ to $\xi = 1$ yields Eq. 4.21:

$$-\int_{\xi=0}^{\xi=1} A(T, p, \xi)\, d\xi = \int_{\xi=0}^{\xi=1} \left(\frac{\partial H}{\partial \xi}\right)_{T,p} d\xi - T\int_{\xi=0}^{\xi=1} \left(\frac{\partial S}{\partial \xi}\right)_{T,p} d\xi. \tag{4.21}$$

By introducing the entropy change $(\Delta S)_{T,p}$ shown in Eq. 4.22 for one equivalent extent of reaction at constant $T$ and $V$:

$$(\Delta S)_{T,p} = \int_{\xi=0}^{\xi=1} \left(\frac{\partial S(\xi)}{\partial \xi}\right)_{T,p} d\xi = -(S_{\xi=1} - S_{\xi=0})_{T,p}, \tag{4.22}$$

we then obtain from Eq. 4.21 the average affinity of the reaction at constant $T$ and $p$ as shown in Eq. 4.23:

$$-A_{T,p}^{\mathrm{mean}} = (\Delta H)_{T,p} - T\,(\Delta S)_{T,p} = Q_{T,p}^{\mathrm{mean}} - T\,(\Delta S)_{T,p}. \tag{4.23}$$

Similarly, we obtain the average affinity of the reaction at constant $T$ and $V$:

$$-A_{T,V}^{\mathrm{mean}} = (\Delta U)_{T,V} - T\,(\Delta S)_{T,V} = Q_{T,V}^{\mathrm{mean}} - T\,(\Delta S)_{T,V}. \tag{4.24}$$

On taking account of Eqs. 4.16, 4.17, 4.19 and 4.20, we rewrite Eqs. 4.23 and 4.24 in the form shown in Eqs 4. 25 and 4.26, respectively:

$$-A_{T,p}^{\mathrm{mean}} = (\Delta G)_{T,p} = (\Delta H)_{T,p} - T\,(\Delta S)_{T,p}, \tag{4.25}$$

$$-A_{T,V}^{\mathrm{mean}} = (\Delta F)_{T,V} = (\Delta U)_{T,V} - T\,(\Delta S)_{T,V}. \tag{4.26}$$

These equations 4.25 and 4.26 are of great importance and are frequently used to estimate the average affinity of a chemical reaction, $A_{T,V}^{\mathrm{mean}} = -(\Delta F)_{T,V}$ or $A_{T,p}^{\mathrm{mean}} = -(\Delta G)_{T,V}$, from the average heat of the reaction, $(\Delta U)_{T,V}$ or $(\Delta H)_{T,p}$, and the entropy changes of the reaction, $(\Delta S)_{T,V}$ or $(\Delta S)_{T,p}$.

Furthermore, from Eq. 4.8 we have Eq. 4.27:

$$\left[\frac{\partial}{\partial T}\left(\frac{A_{T,V}^{\mathrm{mean}}}{T}\right)\right]_V = \frac{(\Delta U)_{T,V}}{T^2}, \qquad \left[\frac{\partial}{\partial T}\left(\frac{A_{T,p}^{\mathrm{mean}}}{T}\right)\right]_p = \frac{(\Delta H)_{T,p}}{T^2}. \tag{4.27}$$

This is the direct relation connecting the average affinity and the average heat of reaction.

Equation 4.27 enables the average affinity to be calculated at a temperature $T$, if its value at some specified temperature $T_0$ is known. If we integrate Eq. 4.27 between $T_0$ and $T$, we have at constant pressure:

$$\frac{A_T^{\text{mean}}}{T} - \frac{A_{T_0}^{\text{mean}}}{T_0} = \int_{T_0}^{T} \frac{(\Delta H)_{T,p}}{T^2}\, dT, \tag{4.28}$$

where the heat of reaction $(\Delta H)_{T,p}$ is a linear function of the partial molar heat capacity of substances taking part in the reaction as shown in Eq. 2.30. We have usually to assume that the partial molar capacities of substances in the reaction system are equal to the molar heat capacities in the pure state. From Eq. 2.30 we have:

$$(\Delta H)_{T,p} = (\Delta H)_{T_0,p} + \int_{T_0}^{T} \sum_i \nu_i\, c_{p,i}\, dT \tag{4.29}$$

which on substitution in Eq. 4. 28 gives Eq. 4.30:

$$\frac{A_T^{\text{mean}}}{T} = \frac{A_{T_0}^{\text{mean}}}{T_0} - \frac{(\Delta H)_{T_0,p}}{T} + \frac{(\Delta H)_{T_0,p}}{T_0} + \sum_i \nu_i \int_{T_0}^{T} \frac{dT}{T^2} \int_{T_0}^{T} c_{p,i}\, dT \tag{4.30}$$

We see that the average affinity $A_T^{\text{mean}}$ of a reaction at a temperature $T$ can be calculated, if we know: (a) the average affinity $A_{T_0}^{\text{mean}}$ at one specified temperature $T_0$ at the pressure $p$; (b) the heat of reaction $(\Delta H)_{T_0,p}$ at $T_0$; and (c) the partial molar heat capacities of the constituent substances as a function of temperature throughout the whole range from $T_0$ to $T$.

At temperature $T_0$, we have from Eq. 4.25, $-A_{T_0,p}^{\text{mean}} = (\Delta H)_{T_0,p} - T_0\,(\Delta S)_{T_0,p}$, which on insertion in Eq. 4.30 gives Eq. 4.31:

$$\frac{A_T^{\text{mean}}}{T} = -\frac{(\Delta H)_{T_0,p}}{T} + (\Delta S)_{T_0,p} + \sum_i \nu_i \int_{T_0}^{T} \frac{dT}{T^2} \int_{T_0}^{T} c_{p,i}\, dT\,. \tag{4.31}$$

The double integral may be transformed by integration by part to Eq. 4.32:

$$\int_{T_0}^{T} \frac{dT}{T^2} \int_{T_0}^{T} c_{p,i}\, dT = \frac{1}{T} \int_{T_0}^{T} dT \int_{T_0} \frac{c_{p,i}}{T}\, dT = \int_{T_0}^{T} \frac{c_{p,i}}{T}\, dT - \frac{1}{T} \int_{T_0}^{T} c_{p,i}\, dT. \tag{4.32}$$

Thus, Eq. 4.31 may be expressed in the alternative form:

$$A_T^{\text{mean}} = -\,(\Delta H)_{T_0,p} + T\,(\Delta S)_{T_0,p} + \sum_i \nu_i \left[ T \int_{T_0}^{T} \frac{c_{p,i}}{T}\, dT - \int_{T_0}^{T} c_{p,i}\, dT \right]. \tag{4.33}$$

With the aid of tables of molar heat capacities in physicochemical handbooks we can evaluate the third term on the right hand side of this equation for each constituent substances taking part in the reaction.

# CHAPTER 5

# CHEMICAL POTENTIAL

The chemical potential is defined as an intensive energy function to represent the energy level of a chemical substance in terms of the partial molar quantity of free enthalpy of the substance. For open systems permeable to heat, work, and chemical substances, the chemical potential can be used more conveniently to describe the state of the systems than the usual extensive energy functions. This chapter discusses the characteristics of the chemical potential of substances in relation with various thermodynamic energy functions. In a mixture of substances the chemical potential of an individual constituent can be expressed in its unitary part and mixing part.

## 5. 1. Thermodynamic Potentials in Open Systems.

We have introduced in the foregoing chapters energy functions (thermodynamic potentials) of extensive properties such as $U$, $H$, $F$, and $G$ to describe the thermodynamic state of a closed system which forbids the exchange of substances with its surroundings. For an open system which allows the exchange of substances to occur with the surroundings, it is often convenient to use energy functions of intensive properties such as the partial molar quantities of energy rather than thermodynamic energy functions of extensive properties.

For a closed system the first law of thermodynamics has defined an energy function called internal energy $U$, which is expressed as a function of the temperature, volume, and number of moles of the constituent substances in the system; $U = U(T, V, n_1 \cdots n_c)$. Furthermore, the second law has defined a state property, called entropy $S$, of the system, which is also expressed as a function of state variables; $S = S(T, V, n_1 \cdots n_c)$. Thermodynamics presumes that the functions $U(T, V, n_1 \cdots n_c)$ and $S(T, V, n_1 \cdots n_c)$ exist independent of whether the system is closed or open. The energy functions of $U$, $H$, $F$, and $G$, then, apply not only to closed systems but also to open systems.

The total differential of the internal energy $U$ of a system can be written as a function of independent state variables such as the temperature, volume and composition of the system as shown in Eq. 5.1:

$$dU = \left(\frac{\partial U}{\partial T}\right)_{V,n_i} dT + \left(\frac{\partial U}{\partial V}\right)_{T,n_i} dV + \sum_i \left(\frac{\partial U}{\partial n_i}\right)_{T,V,n_j} dn_i, \tag{5.1}$$

where $n_i$ is the number of moles of every constituent $i$ and $n_j$ is the number of moles of all constituents $j$ other than $i$. The total differentials of the other energy functions $H$, $F$, and $G$ can also be expressed in the form similar to Eq. 5.1.

**5. 2. The Partial Molar Quantity of Energy and the Chemical Potential.**

We shall now choose $S$, $V$, $n_1 \cdots n_i \cdots$ as independent variables. The internal energy of an open system then yields its total differential $dU$ expressed as a function of these independent variables as shown in Eq. 5.2:

$$dU = \left(\frac{\partial U}{\partial S}\right)_{V,n} dS + \left(\frac{\partial U}{\partial V}\right)_{S,n} dV + \sum_i \left(\frac{\partial U}{\partial n_i}\right)_{S,V,n_j} dn_i \tag{5.2}$$

By substituting $(\partial U/\partial S)_{V,n} = T$ and $(\partial U/\partial V)_{S,n} = -p$ into Eq. 5.2 as shown in Eq. 3.34, we obtain Eq. 5.3 for the total differential of $U$. For all energy functions $U$, $H$, $F$, and $G$, a series of fundamental equations in a form similar to Eq. 5.3 with different characteristic variables can thus be obtained as follows:

$$dU = T\,dS - p\,dV + \sum_i \left(\frac{\partial U}{\partial n_i}\right)_{S,V,n_j} dn_i, \tag{5.3}$$

$$dH = T\,dS + V\,dp + \sum_i \left(\frac{\partial H}{\partial n_i}\right)_{S,V,n_j} dn_i, \tag{5.4}$$

$$dF = -S\,dT - p\,dV + \sum_i \left(\frac{\partial F}{\partial n_i}\right)_{T,V,n_j} dn_i, \tag{5.5}$$

$$dG = -S\,dT + V\,dp + \sum_i \left(\frac{\partial G}{\partial n_i}\right)_{T,V,n_j} dn_i. \tag{5.6}$$

Recalling the definition of these energy functions $H = U + pV$, $F = U - TS$, and $G = U + pV - TS$, we realize that the third terms on the right hand side of these equations are equal to one another as shown in Eq. 5.7:

$$\left(\frac{\partial U}{\partial n_i}\right)_{S,V,n_j} = \left(\frac{\partial H}{\partial n_i}\right)_{S,p,n_j} = \left(\frac{\partial F}{\partial n_i}\right)_{T,V,n_j} = \left(\frac{\partial G}{\partial n_i}\right)_{T,p,n_j} \equiv \mu_i. \tag{5.7}$$

Equation 5.7 defines the *chemical potential* $\mu_i$ of a constituent substance $i$ in the system.

We then obtain a series of the fundamental equations for the total differential of the thermodynamic energy functions as shown in the following equations:

$$dU = T\,dS - p\,dV + \sum_i \mu_i\,dn_i, \tag{5.8}$$

$$dH = T\,dS + V\,dp + \sum_i \mu_i\,dn_i, \tag{5.9}$$

$$dF = -S\,dT - p\,dV + \sum_i \mu_i\,dn_i, \tag{5.10}$$

$$dG = -S\,dT + V\,dp + \sum_i \mu_i\,dn_i. \tag{5.11}$$

Obviously, the chemical potential of a substance is the partial molar quantity of the principal energy functions with respect to the number of moles of the substance at constant values of their respective independent variables in the system as shown in Fig. 5.1.

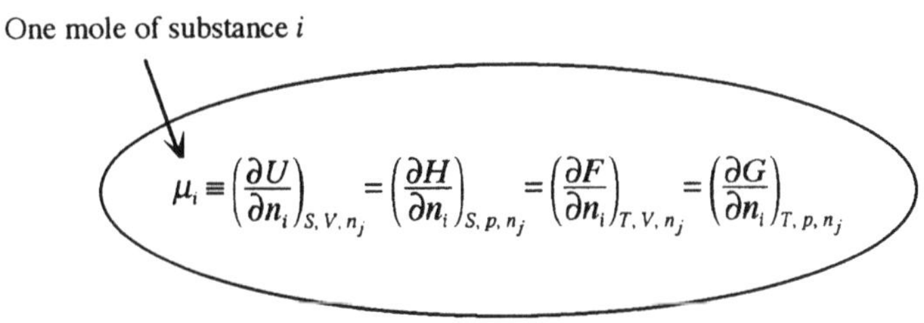


Fig. 5.1. The chemical potential of a substance *i* in a system.

**5. 3. Chemical Potentials and the Affinity of Reaction.**

The affinity of a chemical reaction is in general expressed as a function of the extent of reaction $\xi$ and hence of the number of moles of the chemical substances in the reaction as shown in Eq. 5.12:

$$A = -\left(\frac{\partial G}{\partial \xi}\right)_{T,p} = -\sum_i \nu_i \left(\frac{\partial G}{\partial n_i}\right)_{T,p}, \tag{5.12}$$

where $n_i$ is the number of moles of substance *i* and $\nu_i$ is the stoichiometrical coefficient of *i* in the reaction. We thus obtain Eq. 5.13 for the affinity of the reaction as a function of the chemical potentials of the chemical substances taking part in the reaction:

$$A = -\sum_i \nu_i\,\mu_i. \tag{5.13}$$

This simple form of expression has extensively been used for the calculation of the affinity of chemical reactions.

Consider for instance the reduction of iron oxide to metallic iron by gaseous molecular hydrogen:

$$Fe_2O_{3\,(solid)} + 3\ H_{2\,(gas)} \rightarrow 2\ Fe_{(solid)} + 3\ H_2O_{(gas)}\ .$$

The affinity of this reaction is then given by the following equation:

$$A = \mu_{Fe2O3} + 3\ \mu_{H2} - 2\ \mu_{Fe} - 3\ \mu_{H2O}.$$

As the reaction reaches its equilibrium, the affinity of the reaction decreases to zero as shown in Eq. 5.14:

$$\sum_i \nu_i\, \mu_i = 0, \tag{5.14}$$

Equation 5.14 thus represents the state of equilibrium of the reaction $A = 0$.

The most important property of the chemical potential is that the affinity of a reaction is expressed by the difference in the chemical potential between the reactants and the products as shown in Eq. 5.13 and that the condition of reaction equilibrium is also expressed in terms of the chemical potentials of these reactants and products as shown in Eq. 5.14.

**5. 4 Chemical Potentials and Thermodynamic Energy Functions.**

Among the four principal thermodynamic energy functions, $U$, $H$, $F$, and $G$, the free enthalpy $G$ (Gibbs energy) associated with the intensive variables $T$ and $p$ is a *homogeneous function of the first degree* with respect to the extensive independent variable of the number of moles $n_i$ of the constituent substances present in the system considered, so that it can be expressed as the sum of the chemical potentials of all constituent substances at *constant temperature and pressure*:

$$G = \sum_i n_i\, \mu_i\ . \tag{5.15}$$

From this equation we can derive the other energy functions, $U = G + TS - pV$, $H = G + TS$, and $F = G - pV$, in terms of the chemical potentials of all constituent substances in the system as shown in the following equations:

$$U = \sum_i n_i\, \mu_i + T\, S - p\, V, \tag{5.16}$$

$$H = \sum_i n_i\, \mu_i + T\, S, \tag{5.17}$$

$$F = \sum_i n_i\, \mu_i - p\, V. \tag{5.18}$$

Comparing these four equations from 5.15 to 5.18, we realize that the free enthalpy $G$ (Gibbs energy) is the most convenient in that it is directly proportional to the chemical potentials of the constituent substances and is a function of the characteristic intensive variables $T$ and $p$.

**5. 5. Chemical Potentials in Homogeneous Mixtures: The Gibbs-Duhem Equation.**

From Eq. 5.15 we have Eq. 5.19 for the total differential of the free enthalpy in a homogeneous mixture containing multiple substances:

$$dG = \sum_i n_i \, d\mu_i + \sum_i \mu_i \, dn_i, \tag{5.19}$$

which, on combination with Eq. 5.11, yields Eq. 5.20:

$$S \, dT - V \, dp + \sum_i n_i \, d\mu_i = 0. \tag{5.20}$$

This equation 5.20, called the Gibbs-Duhem equation, is unique among a variety of the thermodynamic equations of state in that the characteristic variables are all intensive quantities, each multiplied by its conjugate extensive quantity.

At a constant temperature and pressure we then obtain Eq. 5.21 for the relation between the chemical potentials and the numbers of moles of the constituent substances:

$$\sum_i n_i \, d\mu_i = 0. \tag{5.21}$$

Equation 5.21 shows the interrelationship among the chemical potentials of the constituent substances in a homogeneous mixture and is often used for the determination of the chemical potential of solute constituents in solutions.

**5. 6. Chemical Potentials of Substances in Ideal Mixtures.**

The chemical potential of a substance $i$ in a homogeneous mixture depends on the temperature, pressure, and concentrations of constituent substances, $\mu_i = \mu_i(T, p, x_1 \cdots x_i \cdots)$; whereas, that of a pure substance is a function of temperature and pressure only. As mentioned in the foregoing chapters, the mixing of substances causes an increase in entropy of the system and hence changes the chemical potentials of the substances

Generally, the chemical potential of a constituent substance $i$ in a mixture consists of *a unitary part*, which is inherent to the pure substance $i$ and independent of its concentration, and *a communal part*, which depends on the concentration of constituent $i$ [Ref. 3.]. The communal part of the chemical potential of a constituent $i$ in a mixture arises from the entropy of mixing of $i$: For an ideal mixture the molar entropy of mixing of $i$, $s_i^M$, is given from Eq. 3.51 by $s_i^M = -R \ln x_i$, and hence the communal part of the chemical potential is expressed by $\mu_i^M = -T s_i^M = RT \ln x_i$ at constant temperature, where $x_i$ is the molar fraction of constituent $i$ in the mixture: $G = H - TS$ gives $\mu_i = h_i - Ts_i$ and hence $\mu_i^M = -Ts_i^M$ for an ideal mixture in which $h_i^M = 0$ (vid. section 5.9.). We then express the chemical potential of substance $i$ in a mixture as shown in Eq. 5. 22:

$$\mu_i(T, p) = \mu_i^*(T, p) + RT \ln x_i, \tag{5.22}$$

where $\mu_i^*(T,p)$ is the unitary part of the chemical potential of $i$ in the mixture. The unitary part $\mu_i^*(T,p)$ is a function of temperature and pressure only. We call $\mu_i^*(T,p)$ the *unitary value* of the chemical potential or simply the *unitary chemical potential* of *i*. The communal part, $RT \ln x_i$, on the right hand side of Eq. 5.22 is called the *chemical potential of mixing*. In general, as described above, the chemical potential of constituent $i$ consists of the *unitary term* $\mu_i^*$ and the *mixing term* $\mu_i^M$ as shown in Eq. 5.23:

$$\mu_i^*(T, p), \quad \text{unitary term.} \qquad \mu_i^M = RT \ln x_i, \quad \text{mixing term.} \tag{5.23}$$

Equation 5.22 holds valid only in the case in which the mixture is an *ideal mixture*. A mixture can thus be called *ideal*, if the chemical potential of the constituent substances, *i*, in the system varies linearly with the logarithm of the molar fraction of *i* at the ratio of *RT*.

If this linear relation between the chemical potential and the logarithm of the molar fraction of *i* holds valid in the whole concentration range extending from $x_i = 0$ to $x_i = 1$, the unitary part of the chemical potential $\mu_i^*(T,p)$ is identical with the chemical potential $\mu_i^0(T,p)$ of the pure substance *i*. The linear relation of Eq. 5.22, however, is not necessarily valid over the whole range of concentrations but in the range of dilute concentrations only. In such a case the unitary part of $\mu_i^*(T,p)$ is usually set at the value estimated by extrapolation from the dilute concentration range to the mole fraction of $x_i = 1$.

Two cases then arise with respect to the ideality of mixtures: One is the case in which the mixture is ideal for all values of $x_i$ and for all constituent substances. This type of mixture is thermodynamically called the *perfect mixture*, for which the *Raoult's law* (a linear relation between $\mu_i$ and $\ln x_i$ in the whole range of concentrations) holds valid and in which the unitary chemical potential $\mu_i^*(T,p)$ of $i$ equals the chemical potential $\mu_i^0(T,p)$ of pure substance *i* for all the substances in the system as shown in Eq. 5.24:

$$\mu_i^*(T, p) = \mu_i^0(T, p). \tag{5.24}$$

The other is the case in which the mixture is ideal when all substances but one (solvent) are at very dilute concentrations. Such mixtures are called *ideal dilute solutions*, for which the *Henry's law* (a linear relation between $\mu_i$ and $\ln x_i$ in a limited range of dilute concentrations) holds valid and in which the equality of Eq. 5.24 is realized only for the main substance present in excess as solvent and not for the solute substances as minor constituents:

$$\mu_i^*(T, p) \neq \mu_i^0(T, p), \qquad \text{for solutes.} \tag{5.25}$$

In this case the unitary value of the chemical potential of solute substance $i$ can be estimated, as mentioned above, by extrapolating the chemical potential of dilute constituent $i$ to $x_i = 1$ from the dilute concentration range in which the linear relation of Eq. 5.22 holds.

If we use the concentration scale of molality $m$ instead of mole fraction $x$, the chemical potential of a solute constituent $i$ is expressed by Eq. 5.26:

$$\mu_i(T, p) = \mu_i^{*m}(T, p) + RT \ln m_i, \tag{5.26}$$

where $\mu_i^{*m}(T, p)$ is called the chemical potential of $i$ at temperature $T$ and pressure $p$ and at the unit concentration in the molality scale. In the case of aqueous dilute solutions we obtain the following relation between the unitary value of chemical potential $\mu_i^{*x}(T, p)$ in the molar fraction scale and that $\mu_i^{*m}(T, p)$ in the molality scale:

$$\mu_i^{*m}(T, p) = \mu_i^{*x}(T, p) - RT \ln 55.51,$$

where 55.51 is the number of moles of water for one kilogram of water.

## 5. 7. Activity and Activity Coefficient.

For a non-ideal mixture in which Eq. 5.22 is not valid, we use a physical quantity called *activity* $a_i$ in place of the molar fraction $x_i$. We then have Eq. 5.27 in place of Eq. 5.22 for the chemical potential of $i$ in a non-ideal mixture:

$$\mu_i(T, p) = \mu_i^{*}(T, p) + RT \ln a_i. \tag{5.27}$$

The second term on the right hand side of this equation, as in the case of Eq. 5.22, represents the communal part of the chemical potential, $\mu_i^M = RT \ln a_i$.

The ratio of the activity $a_i$ to the molar fraction $x_i$ is called the *activity coefficient* $\gamma_i$ :

$$\gamma_i \equiv \frac{a_i}{x_i}. \tag{5.28}$$

Substitution of $a_i$ from Eq. 5.28 in Eq. 5.27 gives Eq. 5.29 for the chemical potential of substance $i$ in the non-ideal mixture:

$$\mu_i(T, p) = \mu_i^{*}(T, p) + RT \ln x_i \gamma_i = \mu_i^{*}(T, p) + RT \ln x_i + RT \ln \gamma_i. \tag{5.29}$$

The third term $RT \ln \gamma_i$ on the right hand side of Eq. 5.29 is called the *excess chemical potential* of $i$.

Chemical thermodynamics also provides the concept of *absolute activity* $\boldsymbol{a}_i$ defined by:

$$\boldsymbol{a}_i = \exp\left(\frac{\mu_i}{RT}\right). \tag{5.30}$$

By contrast, the activity introduced in Eq. 5.27 is the relative activity expressed by Eq. 5.31:

$$a_i = \exp\left(\frac{\mu_i - \mu_i^*}{RT}\right). \tag{5.31}$$

## 5. 8. Chemical Potentials of Pure Substances.

The chemical potential of a pure substance $i$ indicates the thermodynamic energy level of the substance relative to the energy level of the chemical *elements* that make up the substance $i$. In chemical thermodynamics the chemical potentials of elements are conventionally all set zero in the stable state of them at the standard temperature 298 K and pressure 101.3 kPa. The chemical potential of a substance (a chemical compound) $i$ at the standard state, as a result, is equal to the free enthalpy (Gibbs energy) required to form one mole of the substance $i$ from its constituent elements in their stable standard state.

Table 5.1. Standard chemical potentials $\mu_i^0$, standard molar enthalpy $h_i^0$, and standard molar absolute entropy values $s_i^0$ of substances in the standard state of 298 K and 101.3 kP.

| Substance | State | $\mu_i^0/\text{kJ}\cdot\text{mol}^{-1}$ | $h_i^0/\text{kJ}\cdot\text{mol}^{-1}$ | $s_i^0/\text{J}\cdot\text{K}^{-1}\text{mol}^{-1}$ |
|---|---|---|---|---|
| Fe | Solid | 0 | 0 | 27.2 |
| FeO | Solid | - 244.5 | - 266.6 | 54.0 |
| $Fe_3O_4$ | Solid | -741.3 | -822.6 | 90.0 |
| $H_2$ | Gas | 0 | 0 | 130.6 |
| $H_2O$ | Gas | -228.7 | -242.0 | 188.8 |
| $H_2O$ | Liquid | -237.7 | -286.0 | 70.0 |
| $N_2$ | Gas | 0 | 0 | 191.6 |
| C | Solid (Graphite) | 0 | 0 | 5.7 |
| CO | Gas | -137.3 | -110.6 | 198.0 |
| $CO_2$ | Gas | -394.6 | -393.7 | 213.7 |
| $CH_4$ | Gas | -50.80 | -74.88 | 186.4 |
| $CH_3OH$ | Gas | -162.0 | -201.4 | 237.8 |
| $CH_3OH$ | Liquid | -166.4 | -238.8 | 126.8 |

For a chemical reaction the stoichiometrical sum of the chemical potentials of the reactants relative to that of the products is important in thermodynamic examination of the reaction. Let us consider a chemical reaction of solid carbon and gaseous oxygen to form gaseous carbon dioxide:

$$C_{(solid)} + O_{2\,(gas)} \rightarrow CO_{2\,(gas)}. \tag{5.32}$$

From Eqs. 5.13 and 5.22 we obtain the unitary part of the affinity of the reaction as a function of the unitary chemical potentials, $\mu_C^* = \mu_C^{pure}$, $\mu_{O_2}^* = \mu_{O_2}^{pure}$, and $\mu_{CO_2}^* = \mu_{CO_2}^{pure}$, of the reactants and product as shown in Eq. 5.33:

$$A^* = -(\mu_{CO_2}^* - \mu_C^* - \mu_{O_2}^*). \tag{5.33}$$

We call this quantity $A^*$ *the unitary affinity of the reaction*. Since the chemical potentials of solid carbon C and of gaseous molecular oxygen $O_2$ are set zero in the standard state $\left(\mu_C^* = \mu_C^0 = 0,\ \mu_{O_2}^* = \mu_{O_2}^0 = 0\right)$, the unitary affinity of the reaction in the standard state equals minus the *standard chemical potential* of carbon dioxide $\mu_{CO_2}^* = \mu_{CO_2}^{pure}$: $A^* = -\mu_{CO_2}^* = -\mu_{CO_2}^0$. As mentioned above, the standard chemical potential of carbon dioxide $\mu_{CO_2}^0$ is often called the standard *molar free enthalpy of formation* of $CO_2$.

It follows, in general, that the standard chemical potential $\mu_i^0$ of a chemical compound $i$ corresponds to the free enthalpy of formation for one mole of the compound substance $i$ at the standard state, the value of which is tabulated in chemical handbooks as shown for a few compounds in Table 5.1. For ions in electrolytic solutions the chemical potential in their pure state can not be defined, but we may use the standard state of an ion in which the ionic activity is equal to unity ($a_i = 1$) to define the unitary chemical potential of the ion as will be discussed in chapter 9.

## 5. 9. Thermodynamic Potentials in Ideal Mixtures

Starting from the definition 5.22 we now establish several important properties of thermodynamic potentials (partial molar quantities of thermodynamic energy functions) for an ideal system of mixture. Differentiating $G = H - TS$ with respect to $n_i$ with $T$ and $p$ constant, we have $\mu_i = h_i - Ts_i$ and furthermore $[\partial(\mu_i / T) / \partial T]_{p,n_j} = (1/T)(\partial\mu_i/\partial T) - (\mu_i/T^2)$ $= -[(T s_i + \mu_i)/T^2] = -h_i/T^2$. From this equation we obtain Eq. 5.34 for the partial molar enthalpy $h_i^{id}$ of a constituent $i$ in an ideal mixture:

$$-\frac{h_i^{id}}{T^2} = \left[\frac{\partial\left(\frac{\mu_i^{id}}{T}\right)}{\partial T}\right]_{p,j} = \left[\frac{\partial\left(\frac{\mu_i^*}{T}\right)}{\partial T}\right]_{p,j} + \left[\frac{\partial R \ln x_i}{\partial T}\right]_{p,j} = \left[\frac{\partial\left(\frac{\mu_i^*}{T}\right)}{\partial T}\right]_p = -\frac{h_i^*}{T^2}. \tag{5.34}$$

Since the unitary chemical potential $\mu_i^*$ is a function of $T$ and $p$ only, the partial molar enthalpy $h_i^{id}$ of each constituent $i$ of an ideal system is independent of the composition of the system. We also see that in an ideal mixture the partial molar enthalpy $h_i^{\mathrm{id}}$ equals the unitary partial molar enthalpy $h_i^*$ in the whole range of composition.

The partial molar volume $v_i^{\mathrm{id}}$ of the constituents of an ideal system also have this same property as above as shown in Eq. 5.35:

$$v_i^{\mathrm{id}} = \left(\frac{\partial \mu_i^*}{\partial p}\right)_T = v_i^* , \tag{5.35}$$

where $v_i^*$ is the unitary partial molar volume.

The partial molar entropy $s_i^{\mathrm{id}}$, however, depends on the composition as follows:

$$s_i^{\mathrm{id}} = -\left(\frac{\partial \mu_i^*}{\partial T}\right)_p - R \ln x_i = s_i^* - R \ln x_i , \tag{5.36}$$

where $s_i^*$ is the unitary partial molar entropy of constituent $i$ in the ideal system.

From Eqs. 5.34 and 2.27 we obtain the heat of reaction, $(\partial H^{\mathrm{id}} / \partial \xi)_{T,p}$, for a reaction in an ideal system as written in Eq. 5.37:

$$\left(\frac{\partial H^{\mathrm{id}}}{\partial \xi}\right)_{T,p} = -T^2 \sum_i \nu_i \left[\frac{\partial\left(\frac{\mu_i^*}{T}\right)}{\partial T}\right]_p , \tag{5.37}$$

indicating that the heat of reaction in an ideal system is independent of the concentrations and depends only on the temperature and pressure.

Similarly, from Eq. 5.35 and $(\partial V^{\mathrm{id}} / \partial \xi)_{T,p} = \sum_i \nu_i v_i$, the volume change $(\partial V^{\mathrm{id}} / \partial \xi)_{T,p}$ due to the advancement of a reaction, for an ideal system, is given by:

$$\left(\frac{\partial V^{\mathrm{id}}}{\partial \xi}\right)_{T,p} = \sum_i \nu_i \left(\frac{\partial \mu_i^*}{\partial p}\right)_T , \tag{5.38}$$

and hence the volume change due to a reaction is independent of the composition.

Furthermore, from Eq. 5.36 and $(\partial S^{\mathrm{id}} / \partial \xi)_{T,p} = \sum_i \nu_i s_i$, the partial derivative of entropy $(\partial S^{\mathrm{id}} / \partial \xi)_{T,p}$ for a reaction in an ideal system is given by Eq. 5.39:

$$\left(\frac{\partial S^{\mathrm{id}}}{\partial \xi}\right)_{T,p} = -\sum_i \nu_i \left(\frac{\partial \mu_i^*}{\partial T}\right)_T - R \sum_i \nu_i \ln x_i , \tag{5.39}$$

indicating that the entropy change due to a reaction depends on the composition of the system.

## 5. 10. The Unitary and Mixing Terms of Thermodynamic Potentials.

The concept of the unitary and mixing terms described above can apply, in general, not only to the chemical potential of substances in a mixture but also to the other thermodynamic functions such as the partial molar enthalpy $h_i$, the partial molar entropy $s_i$, and the partial molar volume $v_i$ of substances in the mixture. These thermodynamic functions can be derived from the chemical potential $\mu_i$ by partially differentiating it with respect to temperature and pressure. From Eq. 3.37 we have $s_i = \partial\mu_i / \partial T$ and $v_i = \partial\mu_i / \partial p$. From Eq. 5.34 we also have $h_i = -T^2 \, \partial(\mu_i / T) / \partial T$.

For the chemical potential $\mu_i = \mu_i^* + \mu_i^M$ in a non-ideal system;

$$\mu_i^* = \mu_i^*(T, p), \qquad \text{unitary term.} \tag{5.40}$$

$$\mu_i^M = RT \ln x_i \gamma_i, \qquad \text{mixing term.} \tag{5.41}$$

If the system is ideal, then we have $\mu_i^M = RT \ln x_i$.

For the partial molar enthalpy $h_i = h_i^* + h_i^M$ in a non-ideal system:

$$h_i^* = -T^2 \frac{\partial(\mu_i^* / T)}{\partial T}, \qquad \text{unitary term.} \tag{5.42}$$

$$h_i^M = -RT^2 \frac{\partial \ln x_i \gamma_i}{\partial T}, \qquad \text{mixing term.} \tag{5.43}$$

If the system is ideal, we have $h_i^M = 0$ and $h_i = h_i^*$.

For the partial molar entropy $s_i = s_i^* + s_i^M$ in a non-ideal system:

$$s_i^* = -\frac{\partial \mu_i^*}{\partial T}, \qquad \text{unitary term.} \tag{5.44}$$

$$s_i^M = -R \ln x_i \gamma_i - RT \frac{\partial \ln x_i \gamma_i}{\partial T}, \qquad \text{mixing term.} \tag{5.45}$$

If the system is ideal, then we have $s_i^M = -R \ln x_i$.

For the partial molar volume $v_i = v_i^* + v_i^M$ in a non-ideal system:

$$v_i^* = \frac{\partial \mu_i^*}{\partial p}, \qquad \text{unitary term.} \tag{5.46}$$

$$v_i^M = RT \frac{\partial \ln x_i \gamma_i}{\partial p}, \qquad \text{mixing term.} \tag{5.47}$$

If the system is ideal, we have $v_i^M = 0$ and $v_i = v_i^*$.

In conclusion, the partial molar quantity in thermodynamics functions consists of its unitary term and its mixing term as shown above.

# CHAPTER 6

# UNITARY AFFINITY AND EQUILIBRIUM

A chemical reaction proceeds in the direction of decreasing its affinity and reaches equilibrium at which the affinity vanishes. The equilibrium is thus the state at which the unitary affinity of the reaction equals minus the affinity of mixing of the reaction system. The equilibrium constant of a reaction is accordingly an exponential function of the unitary affinity of the reaction. This chapter discusses the role of the unitary affinity in reaction equilibrium

## 6. 1. Affinity and Equilibrium in Chemical Reactions.

Let us consider a chemical reaction in which reactants $R_i$ change into products $P_i$ as shown in Eq. 6.1:

$$\nu_1 \mathrm{R}_1 + \nu_2 \mathrm{R}_2 \rightarrow \nu_3 \mathrm{P}_3 + \nu_4 \mathrm{P}_4, \text{ or } \quad \sum \nu_i \mathrm{R}_i + \sum \nu_i \mathrm{P}_i = 0 \tag{6.1}$$

where $\nu_i$ is the stoichiometrical coefficient. Equations. 5.13 and 5.27 give us the affinity shown in Eq. 6.2:

$$A = -\sum_i \nu_i \mu_i^*(T, p) - \sum_i \nu_i RT \ln a_i = A^* + A^M. \tag{6.2}$$

In the summations, the stoichiometrical coefficient $\nu_i$ is negative for the reactants and positive for the products. In Eq. 6.2 the first term on the right hand side is the *unitary affinity* $A^*$, which comprises of the stoichiometrical sum of the unitary chemical potentials of the reactants and products, and the second term is the *affinity of mixing* $A^M$, which comprises of the stoichiometrical sum of the chemical potentials of mixing for the reactants and products. By substituting the unitary affinity $A^*$ for the first term on the right hand side of Eq. 6.2 and defining this to be equal to $RT \ln K(T, p)$, we obtain Eq. 6.3:

$$A^* = -\sum_i \nu_i \mu_i^*(T, p) \equiv RT \ln K(T, p). \qquad (6.3)$$

Substitution of Eq. 6.3 in Eq. 6.2 then gives Eq. 6.4:

$$A = RT \ln \frac{K(T, p)}{a_1^{\nu_1} a_2^{\nu_2} a_3^{\nu_3} a_4^{\nu_4}}. \qquad (6.4)$$

Since $A = 0$ at equilibrium, Eq. 6.4 yields Eq. 6.5 when the reaction is at equilibrium:

$$K(T, p) = a_1^{\nu_1} a_2^{\nu_2} a_3^{\nu_3} a_4^{\nu_4}. \qquad (6.5)$$

We call $K(T, p)$ the ***equilibrium constant*** of the reaction. As realized from Eq. 6.2, the reaction equilibrium is established at which the unitary affinity $A^*$ becomes equal to minus the affinity of mixing $A^M$ of the reaction.

The unitary affinity of a reaction in the standard state (298 K, 101.3 kPa, and unit activities) is normally called the *standard affinity* $A^0$.

**6. 2. The Unitary Affinity.**

From Eq. 6.3 we have the relation between the unitary affinity $A^*$ and the equilibrium constant $K(T,p)$ of a reaction as shown in Eq. 6.6:

$$K(T, p) = \exp\left[\frac{A^*}{RT}\right]. \qquad (6.6)$$

The unitary affinity of a reaction can be obtained, as mentioned in the foregoing chapter 5, from the unitary chemical potentials of the reactants and products.

Let us consider a reaction of gaseous molecular hydrogen with solid iodine to form gaseous hydrogen iodide as shown in Eq. 6.7:

$$\frac{1}{2} H_{2\,(gas)} + \frac{1}{2} I_{2\,(solid)} \rightarrow HI_{(gas)}. \qquad (6.7)$$

The standard affinity of this reaction at the standard temperature and pressure is expressed by Eq. 6.8:

$$A^0 = \frac{1}{2} \mu_{H_2}^0 + \frac{1}{2} \mu_{I_2}^0 - \mu_{HI}^0. \qquad (6.8)$$

With $\mu_{H_2}^0 = 0$, $\mu_{I_2}^0 = 0$, and $\mu_{HI}^0 = -1.3\ \mathrm{kJ \cdot mol^{-1}}$ found in chemical handbooks, we then obtain the standard affinity equal to $A^0 = -\mu_{HI}^0 = 1.3\ \mathrm{kJ \cdot mol^{-1}}$. If the gas phase is an ideal gas, the activities are equal to the molar fraction of gaseous substances. Further, if solid $I_2$ is a pure substance, the activity of $I_2$ is unity. The equilibrium constant $K$ in the standard state will then be expressed by the molar fractions of gaseous constituents as shown in Eq. 6.9:

$$K = \frac{x_{\mathrm{HI}}}{x_{\mathrm{H_2}}^{1/2}} = 0.593, \tag{6.9}$$

Equation 6.9 gives the molar fractions of gaseous molecular hydrogen and hydrogen iodide $x_{\mathrm{H_2}} = 0.558$ and $x_{\mathrm{HI}} = 0.442$, respectively, in the reaction equilibrium at the standard state.

## 6. 3. Equilibrium Constants and Concentration Units.

The equilibrium constant $K$ of a reaction is dimensionless but we can express concentration in different units. For a gaseous mixture, in addition to the molar fraction $x_i$, two other concentration units may be used: One is the partial pressure $p_i = x_i p$, which is proportional to the molar fraction $x_i$ and the total pressure $p$; and the other is the *molar concentration* (molarity) $c_i = n_i / V$, which is inversely proportional to the volume $V$ of the gaseous mixture. In terms of these concentration units the equilibrium constant of a gas reaction is expressed in three different formulas shown, respectively, in Eq. 6.10:

$$K_x(T, p) = x_1^{\nu_1}\, x_2^{\nu_2}\, x_3^{\nu_3}\, x_4^{\nu_4}, \quad K_p(T) = p_1^{\nu_1}\, p_2^{\nu_2}\, p_3^{\nu_3}\, p_4^{\nu_4}, \quad K_c(T) = c_1^{\nu_1}\, c_2^{\nu_2}\, c_3^{\nu_3}\, c_4^{\nu_4}. \tag{6.10}$$

For an ideal gas in which the equation of state $p_i = c_i RT$ holds valid, we have Eqs. 6.11, 6.12, and 6.13 as the relation of the unitary affinity $A^*$ with either of the equilibrium constants, $K_x(T, p)$, $K_p(T)$, and $K_c(T)$, expressed in the three different concentration units:

$$A^* = RT \ln K_x(T, p), \tag{6.11}$$

$$A^* = RT \ln K_p(T) - \nu\, RT \ln p, \tag{6.12}$$

$$A^* = RT \ln K_c(T) - \nu\, RT \ln p + \nu\, RT \ln RT, \tag{6.13}$$

where $\nu = \Sigma_i \nu_i$ is the sum of the stoichiometrical coefficients of the reactants and products. We hence obtain the relationship among the three different expressions of the equilibrium constant of the reaction, $K_x$, $K_p$, and $K_c$, as shown in Eqs. 7.14 and 7. 15:

$$K_x(T, p) = p^{-\nu}\, K_p(T), \tag{6.14}$$

$$K_c(T) = [RT]^{-\nu}\, K_p(T). \tag{6.15}$$

Let us take, for example, a reaction decomposing water vapor into gaseous molecular hydrogen and oxygen in which $\nu_{\mathrm{H_2O}} = -2$, $\nu_{\mathrm{H_2}} = +2$, and $\nu_{\mathrm{O_2}} = +1$ as shown in Eq. 6.16:

$$2\,\mathrm{H_2O_{(gas)}} \rightarrow 2\,\mathrm{H_{2,(gas)}} + \mathrm{O_{2,(gas)}}\,. \tag{6.16}$$

Since $\nu = \Sigma_i \nu_i = +1$, we obtain Eq. 6.17:

$$K_x(T,p) = \frac{K_p(T)}{p} = \frac{RT}{p} K_c(T), \tag{6.17}$$

which shows that $K_x(T,p)$ is inversely proportional to the pressure $p$, while $K_p(T)$ and $K_c(T)$ are independent of the pressure $p$ of the system. It is then advantageous to use $K_p(T)$ and $K_c(T)$ rather than $K_x(T,p)$ for the reactions involving gaseous substances.

## 6. 4. Equilibrium Constants as a Function of Pressure and Temperature.

We now discuss the partial differential with respect to pressure $p$ and temperature $T$ of the equilibrium constant $K_x(T,p)$ in terms of the molar fraction. From the definition in Eq. 6.3 we obtain immediately Eq. 6.18 for the partial differential of the logarithm of $K_x$ with respect to pressure $p$:

$$\left(\frac{\partial \ln K_x(T,p)}{\partial p}\right)_T = -\frac{1}{RT}\sum_i \nu_i \left(\frac{\partial \mu_i^*(T,p)}{\partial p}\right)_T, \tag{6.18}$$

where $\nu_i$ is the stoichiometrical coefficient of $i$ in the reaction. From Eq. 5.46 $v_i^* = \partial\mu_i/\partial p$, the right hand side of Eq. 6.18 contains the stoichiometrical sum of the unitary partial molar volume $v_i^*$ of the reactants and products: $\sum_i \nu_i v_i^*$. For an ideal reaction system $v_i^*$ is equal to the partial molar volume $v_i$ of substance $i$: $\sum_i \nu_i v_i^* = \sum_i \nu_i v_i = (\partial V/\partial\xi)_{T,p}$, where $V$ is the volume of the reaction system.

In the case of an ideal reaction system therefore we have Eq. 6.19:

$$\left(\frac{\partial V}{\partial \xi}\right)_{T,p} = \sum_i \nu_i v_i = \sum_i \nu_i \left(\frac{\partial \mu_i^*(T,p)}{\partial p}\right)_T, \tag{6.19}$$

which can also be derived from Eqs. 3.38 and 5.37. We hence obtain Eq. 6.20 for the pressure-dependence of the reaction equilibrium constant:

$$\left(\frac{\partial \ln K_x(T,p)}{\partial p}\right)_T = -\frac{1}{RT}\left(\frac{\partial V}{\partial \xi}\right)_{T,p} \tag{6.20}$$

Thus the partial differential of the logarithm of $K_x$ with respect to pressure $p$ is equal to minus the molar expansion, $\partial V/\partial\xi$, divided by $RT$. We see then that an increase in pressure increases the equilibrium constant $K_x$ if the reaction is accompanied by a decrease in volume ($\partial V/\partial\xi < 0$), and conversely if $\partial V/\partial\xi > 0$ the equilibrium constant is decreased.

Similarly, the partial differential of $\ln K_x$ with respect to temperature is given by Eq. 6.21:

$$\left(\frac{\partial \ln K_x(T,p)}{\partial T}\right)_p = -\frac{1}{RT^2}\left(\frac{\partial H}{\partial \xi}\right)_{T,p}. \tag{6.21}$$

which can be derived from Eqs. 6.3, 5.36, and 4.10 for an ideal reaction system in which the mixing term of the partial molar enthalpy $h_i^M$ is zero. Thus the partial differential of the logarithm of $K_x$ with respect to temperature $T$ is equal to minus the heat of reaction $\partial H/\partial\xi$ divided by $RT^2$. If the reaction is accompanied by an absorption of heat ($\partial H/\partial\xi > 0$) the equilibrium constant increases with increasing temperature, whereas for an exothermic reaction ($\partial H/\partial\xi < 0$) it decreases with increasing temperature.

By integrating Eq. 6.21 and assuming that $(\partial H/\partial\xi)_{T,p}$ is independent of $T$, we obtain for $K_p(T)$ Eq. 6.22:

$$\ln K_p(T) = -\frac{1}{RT}\left(\frac{\partial H}{\partial \xi}\right)_{T,p} + C, \tag{6.22}$$

where $C$ is an integral constant. This equation, called *van't Hoff's equation*, indicates that a linear relation between $\ln K_p(T)$ and the reciprocal absolute temperature $1/T$ holds as shown in Fig. 6.1. The slope of van't Hoff's plot can be used for an estimation of the heat of reaction $(\partial H/\partial\xi)_{T,p}$.

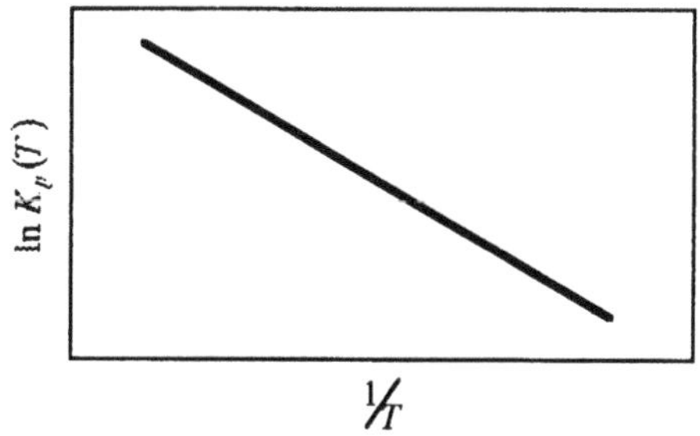


Fig. 6.1. van't Hoff's linear relation between the equilibrium constant and the reciprocal absolute temperature for a chemical reaction.

# CHAPTER 7

# GASES, LIQUIDS, AND SOLIDS

A gaseous substance at dilute density normally is in the state of an ideal gas and it turns into a non-ideal gas as the density increases. A further increase in the density leads to the condensation of a gas into a liquid or solid phase. In the ideal gaseous state the chemical potential of a substance changes linearly with the logarithm of the density, and a deviation from the linearity occurs in the non-ideal state. For a condensed substance in the liquid or solid state its chemical potential hardly changes with the density. This chapter concerns the equations of state and the calculation of thermodynamic potentials of gaseous and condensed substances.

## 7. 1. Perfect and Ideal Gases.

For an ideal gas the internal energy $U$ depends on the temperature $T$ only (Joule's law) and the volume $V$ is inversely proportional with the pressure $p$ at constant temperature (Boyle's law). Equation 7.1 shows the equation of state for an ideal gas:

$$p\,V = n\,RT, \tag{7.1}$$

where $n$ is the number of moles in the gas and $R$ is the gas constant ( $R = 8.314\,\mathrm{J \cdot K^{-1} \cdot mol^{-1}}$). The gas for which Eq. 7.1 holds is called the *perfect gas* or the *ideal gas:* The two terms perfect and ideal mean the same for gases, but they do not mean the same for liquid solutions as will be mentioned in chapter 8.

We shall first consider a perfect or ideal gas of a single substance and discuss its molar enthalpy, molar entropy, and chemical potential as a function of temperature and pressure. From Eqs. 2.31 and 2.32 we obtain the molar enthalpy $h$ of an ideal gas as shown in Eq. 7.2:

$$h(T) = h(T^0) + \int_{T^0}^{T} c_p(T)\,dT, \tag{7.2}$$

where $h(T^0)$ is the molar enthalpy at a reference temperature $T^0$ (the standard temperature) and $c_p(T)$ is the molar heat capacity of the ideal gas: the molar enthalpy of an ideal gas is independent of pressure as shown in Eq. 2.32. We further obtain Eq. 7.3 for the molar entropy $s$ from its differential $ds = (dh - vdp)/T = (c_p/T)dT - (v/T)dp$ shown in Eq. 3.41:

$$s(T, p) = \left[ s(T^0, p^0) + \int_{T^0}^{T} \frac{c_p(T)}{T} \, dT \right] - R \ln \frac{p}{p^0} = s(T, p^0) - R \ln \frac{p}{p^0}. \tag{7.3}$$

Equations 7.2 and 7.3 then give us the chemical potential $\mu = h - Ts$ as follows:

$$\mu(T, p) = \left[ h(T^0) - Ts(T^0, p^0) + \int_{T^0}^{T} c_p(T) \, dT - T \int_{T^0}^{T} \frac{c_p(T)}{T} \, dT \right] + RT \ln \frac{p}{p^0}, \tag{7.4}$$

which is also expressed as Eq. 7.5:

$$\mu(T, p) = \mu(T, p^0) + RT \ln \frac{p}{p^0}, \tag{7.5}$$

where $s(T, p^0)$ and $\mu(T, p^0)$ are the molar entropy and the chemical potential of an ideal gas at temperature $T$ and at the standard pressure $p^0$, respectively. Equation 7.4 is frequently used for calculating the chemical potential of a gaseous substance from its molar heat capacity $c_p(T)$.

With an ideal gas mixture we also have the same equation as Eq. 7.5 for the chemical potential $\mu_i(T, p)$ of one of the constituent substances, $i$, in the ideal gas mixture at the total pressure $p$ and at temperature $T$ as shown in Eq. 7.6:

$$\mu_i(T, p_i) = \mu_i^{\mathrm{pure}}(T, p_i^0) + RT \ln \frac{p_i}{p_i^0}, \tag{7.6}$$

where $p_i$ is the partial pressure of $i$, and $\mu_i^{\mathrm{pure}}(T, p_i^0)$ is the chemical potential of pure substance $i$ at the standard pressure $p_i^0$. The chemical potential $\mu_i(T, p, x)$ in terms of the molar fraction $x_i = p_i/p$ is then obtained as follows:

$$\mu_i(T, p, x) = \mu_i^*(T, p) + RT \ln x_i, \tag{7.7}$$

$$\mu_i^*(T, p) = \mu_i^{\mathrm{pure}}(T, p^0) + RT \ln \left( \frac{p}{p^0} \right), \tag{7.8}$$

where the first term $\mu_i^*(T, p)$ on the right hand side of Eq. 7.7 shows the *unitary* chemical potential of $i$ at temperature $T$ and total pressure $p$, and the first term $\mu_i^{\mathrm{pure}}(T, p^0)$ on the right hand side of Eq. 7.8 is the chemical potential of pure gaseous substance $i$ at the standard pressure $p^0$ and at temperature $T$. The chemical potential of pure substance $i$, $\mu_i^{\mathrm{pure}}(T^0, p^0)$, at the standard temperature $T^0$ and pressure $p^0$ is called the standard chemical potential.

**7. 2. Non-ideal Gases.**

Real gases are usually non-ideal. Thermodynamics describes both ideal and non-ideal gases with the same type of formulas, except that for non-ideal gas mixtures the *fugacity* $f_i$ is substituted in place of the pressure $p_i$ and that the *activity* $a_i$ is substituted in place of the molar fraction $x_i$ or concentration $c_i$ of constituent substance $i$. We have already seen that in the ideal gas of a pure substance the chemical potential is expressed by Eq. 7.5. By analogy, we write Eq. 7.9 for the non-ideal gas of a pure substance $i$:

$$\mu(T, p) = \mu^0(T, f^0) + RT \ln \frac{f(T, p)}{f^0}, \tag{7.9}$$

where $\mu^0(T, f^0)$ is the chemical potential of $i$ at the standard fugacity $f^0$ and at temperature $T$. The fugacity $f(T,p)$ is a function of $T$ and $p$, and it approaches $p$ as the pressure decreases toward zero, i.e. as the state of the gas approaches the ideal gas state:

$$\lim_{p \to 0} \frac{f(T, p)}{p} = 1. \tag{7.10}$$

This ratio of $f$ to $p$ for a non-ideal gas of a pure substance may be calculated from the equation of state for real gases such as the *virial equation* and the *van der Waals equation.*

In a mixture of non-ideal gases, similarly, the chemical potential $\mu_i(T, p, p_i)$ of one of the constituent substances $i$ is expressed by Eq. 7.11 in the same form as Eq. 7.6:

$$\mu_i(T, p, p_i) = \mu_i^*(T, f_i^0) + RT \ln \frac{f_i(T, p, p_i)}{f_i^0}, \tag{7.11}$$

where we have substituted the fugacity $f_i(T, p, p_i)$ for the partial pressure $p_i$ in Eq. 7.6. The term $\mu_i^*(T, f_i^0)$ is the *unitary* chemical potential of substance $i$ in the gas mixture at the standard fugacity $f_i^0$ and at temperature $T$.

If we use instead of the partial pressure $p_i$ the molar fraction $x_i$ to express the concentration of $i$, the activity $a_i$ can be substituted for $x_i$ to obtain Eq. 7.12 for the chemical potential of $i$ in a non-ideal gas mixture:

$$\mu_i(T, p, x) = \mu^*(T, p, a_i^0) + RT \ln \frac{a_i(T, p, x)}{a_i^0}, \tag{7.12}$$

where the term $\mu^*(T, p, a_i^0)$ is the *unitary* chemical potential of $i$ at the standard unit activity $a_i^0 = 1$ at the total pressure $p$ and at temperature $T$. Equation 7.12 for a non-ideal gas mixture is equivalent to Eq. 7.7 for an ideal gas mixture. If the pressure of the gas mixture approaches zero, the activity $a_i$ approaches the molar fraction $x_i$:

$$\lim_{p \to 0} \frac{a_i(T, p, x)}{x_i} = 1. \tag{7.13}$$

The ratio of the activity $a_i$ to the molar fraction $x_i$ is called the *activity coefficient* $\gamma_i$:

$$\gamma_i = \frac{a_i}{x_i}. \tag{7.14}$$

We also define the fugacity coefficient $\varphi_i$ as the ratio of fugacity $f_i$ to partial pressure $p_i$:

$$\varphi_i = \frac{f_i}{p_i}, \tag{7.15}$$

whose value is not the same as that of the activity coefficient $\gamma_i$.

**7. 3. Liquids and Solids.**

Liquids and solids are in the condensed state in which chemical substances are very dense and hardly undergo any volume change with changing pressure in the range of ordinary pressures. Let us now consider a condensed system of a pure substance. The *coefficient of thermal expansion* $\alpha$ and the *compressibility* $\kappa$ are defined in terms of the molar volume $v$ by the following two equations, respectively:

$$\alpha = \frac{1}{v}\left(\frac{\partial v}{\partial T}\right)_p, \qquad \kappa = -\frac{1}{v}\left(\frac{\partial v}{\partial p}\right)_T. \tag{7.16}$$

These coefficients in general are extremely small; $\alpha \approx 10^{-5}\mathrm{K}^{-1}$ and $\kappa \approx 10^{-11}\mathrm{Pa}^{-1}$ for metallic copper at room temperature. The compressibility $\kappa$ is one hundred times as large for liquids as that for solids and is some thousand times as large for gases as that for liquids.

Besides $\alpha$ and $\kappa$, we also make use of the *pressure coefficient* $\beta$ defined by Eq. 7.17:

$$\beta = \frac{1}{p}\left(\frac{\partial p}{\partial T}\right)_V. \tag{7.17}$$

From the mathematical identity $(\partial v / \partial T)_p + (\partial v / \partial p)_T(\partial p / \partial T)_V = 0$ (the partial differential of an implicit function), we obtain the relationship between these three coefficients as shown in Eq. 7.18:

$$\beta = \frac{1}{p}\frac{\alpha}{\kappa}, \tag{7.18}$$

which enables us to calculate $\beta$ if the values of $\alpha$ and $\kappa$ are known..

Normally, the coefficient of thermal expansion $\alpha$ of a solid approaches a certain constant value at high temperatures and falls steeply as the temperature is lowered. This follows from $(\partial V/\partial T)_p = -(\partial S/\partial p)_T$ obtained by the differentiation of $V$ and $-S$ in Eq. 3.37. The third law of thermodynamics assumes that the entropy $S$ falls toward zero as the temperature approaches zero in the absolute temperature scale, and hence both $(\partial S/\partial p)_T$ and $(\partial V/\partial T)_p$ must be close to zero at sufficiently low temperatures.

The coefficient of thermal expansion $\alpha$ of a condensed substance is related to the molar heat capacities $c_p$ at constant pressure. The above equation $(\partial v/\partial T)_p = -(\partial s/\partial p)_T$ for one mole can be differentiated with respect to $T$ and combined with $(\partial s/\partial T)_p = c_p/T$ to obtain the following equation:

$$\left(\frac{\partial^2 v}{\partial T^2}\right)_p = -\left(\frac{\partial^2 s}{\partial T\,\partial p}\right) = -\frac{1}{T}\left(\frac{\partial c_p}{\partial p}\right)_T, \qquad \left(\frac{\partial c_p}{\partial p}\right)_T = -T\left(\frac{\partial^2 v}{\partial T^2}\right)_p. \tag{7.19}$$

Then, Eq. 7.16 give us Eq. 7.20 [Ref. 1.]:

$$\left(\frac{\partial c_p}{\partial p}\right)_T = -Tv\left(\frac{\partial \alpha}{\partial T} + \alpha^2\right) \approx -Tv\,\frac{\partial \alpha}{\partial T}, \tag{7.20}$$

where $\alpha^2$ is negligibly small compared with $\partial\alpha/\partial T$. The variation of the molar heat capacity $c_p$ with pressure is thus related to the variation of $\alpha$ with temperature. Since $\partial\alpha/\partial T$ has always a positive sign, $c_p$ decreases with increased pressure, but the effect is known to become small at high temperatures since $\partial\alpha/\partial T$ falls off more rapidly than the temperature $T$ increases [Ref. 1.].

## 7. 4. The State Equation and Thermodynamic Functions of Condensed Substances.

In the case where the external pressure is not too high (< 100 atm.) and the compressibility remains independent of pressure, the second equation of 7.16 can be integrated with respect to pressure $p$ to obtain Eq. 7.21 for the molar volume $v(T, p)$ of a condensed substance:

$$v(T, p) = v(T, 0)\exp(-\kappa p) \approx v(T, 0)(1 - \kappa p), \tag{7.21}$$

where $v(T, 0)$ is the molar volume extrapolated to zero pressure at constant temperature $T$. Equation 7.21 is the equation of state of a condensed substance at ordinary pressures and remains valid provided that $\kappa p << 1$. If we extend the equation of state to cover a wide range of pressures, a series expansion of the equation may be made as shown in Eq. 7.22:

$$v(T, p) = v(0, 0)(1 + a_0 + a_1 p + \cdots), \tag{7.22}$$

where $v(0,0)$ is the molar volume extrapolated to $T = 0$ and $p = 0$, and $a_0$, $a_1, \cdots$ are functions of temperature only. The volume $v(T,0)$ in Eq. 7.21 is for zero pressure $p = 0$ at a temperature $T$ so that it is expressed by $v(T, 0) = v(0, 0)(1 + a_0)$, where $a_0(T)$ must be zero when $T = 0$. At ordinary temperatures $a_0(T)$ is very small compared with one because the coefficient of thermal expansion is very small. We thus obtain from Eq. 7.21 with $v(T,0)$ given above the *equation of state* of a condensed substance as shown in Eq. 7.23:

$$v(T, p) \approx v(0, 0)(1 + a_0)(1 - \kappa p) \tag{7.23}$$

We next discuss the molar enthalpy $h$, the molar entropy $s$, and the chemical potential $\mu$ of a condensed pure substance as a function of temperature and pressure. For molar enthalpy $h$ Eqs. 2.15 and 3.37 yield Eq. 7.24 through a little complicated derivation [Ref. 1.]:

$$\left(\frac{\partial h}{\partial T}\right)_p = c_p, \quad \left(\frac{\partial h}{\partial p}\right)_T = v - T\left(\frac{\partial v}{\partial T}\right)_p. \tag{7.24}$$

In this derivation Eq. 3.37 gives $(\partial s/\partial T)_{p,\xi} = -(\partial v/\partial p)_{T,\xi}$, Eq. 2.15 gives the latent heat of pressure change for one mole $h_{T,\xi} = (\partial h/\partial p)_{T\xi} - v$, and $h_{T,\xi}$ is also given by $h_{T,\xi} = T(\partial s/\partial p)_{T\xi}$ from the definition of entropy. These equations yield Eq. 7.24. As mentioned in section 2.6, $(\partial h/\partial p)_T = 0$ in the case of an ideal gas, since $v = T\,(\partial v/\partial T)_p$.

Equation 7.24 provides the dependence of temperature and pressure on the molar enthalpy. The molar enthalpy $h$ is then obtained upon integration of Eq. 7.24 with respect to $T$ and $p$ to give Eq. 7.25 equivalent to Eq. 2.31:

$$h(T,p) = h(0,0) + \int_0^T c_p(T,0)\,dT + \int_0^p \left\{v - T\left(\frac{\partial v}{\partial T}\right)_p\right\} dp, \tag{7.25}$$

where $c_p(T,0)$ is the molar heat capacity extrapolated to $p = 0$ and $h\,(0,0)$ is the molar enthalpy extrapolated to $T = 0$ and $p = 0$. In Eq. 7.25 the second term is the thermal part of enthalpy and the third term is the pressure-dependent part of enthalpy. Taking Eqs. 7.16 and 7.23 into account, we obtain the pressure-dependent part of enthalpy as follows:

$$\int_0^p \left\{v - T\left(\frac{\partial v}{\partial T}\right)_p\right\} dp = \int_0^p v\,(1-\alpha T)\,dp = pv(T,0)\,(1-\alpha T)\left(1 - \frac{1}{2}\kappa p\right). \tag{7.26}$$

The molar enthalpy $h(T,p)$ is thus given by Eq. 7.27:

$$h(T,p) = h(0,0) + \int_0^T c_p(T,0)\,dT + pv(T,0)\,(1-\alpha T)\left(1 - \frac{1}{2}\kappa p\right) \tag{7.27}$$

Since $\kappa p \ll 1$ at ordinary temperatures, the molar enthalpy $h(T,p)$ varies linearly with the pressure and the magnitude of the variation is given by the term $pv(T,0)$, which is usually very small compared with the first and second terms in Eq. 7.27. To most purposes, then, the enthalpy may be taken as independent of the pressure.

For the molar entropy of a condensed substance, we obtain Eq. 7.28 from $(\partial s/\partial T)_p = c_p/T$ and $(\partial s/\partial p)_T = -(\partial v/\partial T)_p$ of Eq. 3.37:

$$s(T,p) = s(0,0) + \int_0^T \frac{c_p(T,0)}{T}\,dT - \int_0^p \left(\frac{\partial v}{\partial T}\right) dp. \tag{7.28}$$

where $s(0,0)$ extrapolated to $T = 0$ and $p = 0$ is zero based on the third law of thermodynamics. Substituting Eq. 7.23 in Eq. 7.28 and assuming $\alpha$ to be independent of pressure, we obtain:

$$s(T, p) = s(0, 0) + \int_0^T \frac{c_p(T, 0)}{T} dT - \alpha p v(T, 0)\left(1 - \frac{1}{2}\kappa p\right), \tag{7.29}$$

which indicates that the molar entropy decreases linearly with increasing pressure. The third term in Eq. 7.29, however, is very small compared with the second term so that we may regard the molar entropy $s$ of liquids or solids as independent of the pressure.

$$s(T, p) \approx s(0, 0) + \int_0^T \frac{c_p(T, 0)}{T} dT, \quad \text{condensed substances.} \tag{7.30}$$

Finally, for chemical potential $\mu = h - Ts$ we have Eq. 7.31:

$$\mu(T, p) = h(0, 0) - Ts(0, 0) + \int_0^T c_p(T, 0)\, dT - T\int_0^T \frac{c_p(T, 0)}{T} dT$$
$$+ pv(T, 0)\left(1 - \frac{1}{2}\kappa p\right), \tag{7.31}$$

which can also be expressed by replacing the two integrals with a double integral as follows:

$$\mu(T, p) = \left[h(0, 0) - Ts(0, 0) - \int_0^T \frac{dT}{T^2}\int_0^T c_p(T, 0)\, dT\right] + pv(T, 0)\left(1 - \frac{1}{2}\kappa p\right). \tag{7.32}$$

We thus have the chemical potential $\mu(T, p)$ of a condensed substance at temperature $T$ and pressure $p$ in the formula shown in Eq. 7.33:

$$\mu(T, p) = \mu(T, 0) + pv(T, 0)\left(1 - \frac{1}{2}\kappa p\right), \tag{7.33}$$

where $\mu(T, 0)$ is the chemical potential extrapolated to $p = 0$ at temperature $T$. Normally, $(1 - 0.5\kappa p) \approx 1$. The chemical potential $\mu(\textit{liquid or solid})$ of a liquid or solid substance is thus seen to change linearly with the product of $p$ and $v$ in contrast with the chemical potential $\mu(gas)$ of a gas which varies linearly with the logarithm of $p$ as shown in Eq. 7.5.

Since the second term $pv(T, 0)(1 - 0.5\kappa p)$ on the right hand side of Eq. 7.33 is negligibly small at ordinary pressures, the chemical potential of a condensed substance hardly depends on the pressure of the system:

$$\mu(T, p) \approx \mu(T), \quad \text{condensed substances.} \tag{7.34}$$

# CHAPTER 8

# SOLUTIONS

Solutions are thermodynamically classified into perfect, ideal, and non-ideal solutions. This chapter discusses the characteristics of these solutions and define the excess functions of non-ideal solutions. Also examined are electrolytic solutions which contain dissociated ions.

## 8. 1. Ideal and Non-ideal Solutions.

A solution is defined as a condensed phase (liquid or solid) containing several substances. The main substance of the solution is called *solvent* and the other constituent substances dissolved in the solvent are *solutes*. Solutions are classified into *ideal* solutions and *non-ideal* solutions. For an ideal solution the chemical potential of a constituent substance $i$ is given by:

$$\mu_i = \mu_i^{*}(T, p) + RT \ln x_i \,. \qquad (8.1)$$

From Eq. 7.32 we have the *unitary* chemical potential $\mu_i^{*}(T,p)$ as follows:

$$\mu_i^{*}(T, p) = \mu_i^{**}(T, 0) + p\, v_i(T, 0)\left(1 - \tfrac{1}{2}\, \kappa_i\, p\right), \qquad (8.2)$$

where $\mu_i^{**}(T,0)$ is the unitary chemical potential of constituent $i$ at temperature $T$ and at zero pressure, and $v_i(T,0)$ is the partial molar volume of $i$ extrapolated to zero pressure. A parameter $\kappa_i$ in Eq. 8.2 is defined as a coefficient of compressibility, $\kappa_i = -(1/v_i)(\partial v_i/\partial p)_T$, for each constituent substance $i$ in the ideal solution.

Equation 8.1 is valid for ideal solutions only. In the case of non-ideal concentrated solutions, an activity coefficient $\gamma_i$ is inserted as an adjusting coefficient to keep the expression of chemical potential $\mu_i$ in the same form as Eq. 8.1:

$$\mu_i = \mu_i^{*}(T, p) + RT \ln x_i\, \gamma_i \,. \qquad (8.3)$$

We bear in mind however that the values of $\mu_i^*(T,p)$ and $\gamma_i$ depend upon the choice of the ideal reference system. If we choose for the solvent a reference system in which $\gamma_i$ becomes unity as $x_i$ approaches unity, the unitary chemical potential $\mu_i^*(T,p)$ is given by the chemical potential $\mu_i^0(T,p)$ of the pure solvent *i*: $\mu_i^*(T,p) = \mu_i^0(T,p)$. On the other hand, if we choose for the solute substances a reference system in which $\gamma_i$ becomes unity as $x_i$ approaches zero, the unitary chemical potential $\mu_i^*(T,p)$ is given by the chemical potential $\mu_i^\infty(T,p)$ of the solute *i* at infinite dilution: $\mu_i^*(T,p) = \mu_i^\infty(T,p)$.

Instead of characterizing the deviation from ideality for the solvent 1 in terms of its activity coefficient $\gamma_1$, we may introduce the *osmotic coefficient* $\phi$ defined by $\phi = (\ln x_1\gamma_1)/\ln x_1$ as shown in Eq. 8.4:

$$\mu_1 = \mu_1^*(T, p) + \phi\, RT \ln x_1. \tag{8.4}$$

Since $\ln x_1 = \ln(1 - \Sigma_i x_i) \approx -\Sigma_i x_i$ when $\Sigma_i x_i < 1$, where $x_i$ is the molar fraction of solute *i*, we have an approximate equation for Eq. 8.4 as follows:

$$\mu_1 = \mu_1^*(T, p) - \phi\, RT \sum x_i. \tag{8.5}$$

We then lose some of the formal resemblance to Eq. 8.1 of ideal solutions, but on the other hand the use of $\phi$ is advantageous in that it is much more sensitive to characterize the deviation from ideality than $\gamma_1$. The osmotic coefficient $\phi$ is, in fact, the same coefficient as what is called the boiling or freezing point coefficient.

### 8. 2. Perfect Solutions and Ideal Solutions.

A solution is called *perfect*, if Eq. 8.1 is valid over the whole range of concentration for all constituent substances. The perfect solution is realized if the molecules of the solvent and the solutes are similar to one another in their nature. In perfect solutions the unitary chemical potential $\mu_i^*(T,p)$ of a constituent substance *i* equals the chemical potential $\mu_i^0(T,p)$ of the pure substance *i* for all the constituent substances: Raoult's law.

Let us consider for simplification a binary perfect solution consisting of solvent 1 and solute 2. The free enthalpy (Gibbs energy) for one mole of a binary mixture $g_{\text{mixture}}$ is then given by Eq. 8.6:

$$g_{\text{mixture}} = x_1\,\mu_1 + x_2\,\mu_2 = g^0_{\text{mean}} + RT\,x_1 \ln x_1 + RT\,x_2 \ln x_2, \tag{8.6}$$

where $g^0_{\text{mean}}$ is the mean molar free enthalpy prior to the mixing of $n_1$ moles of solvent 1 and $n_2$ moles of solute 2; that is $g^0_{\text{mean}} = (n_1 g_1^0 + n_2 g_2^0)/(n_1 + n_2) = x_1 g_1^0 + x_2 g_2^0$. The molar free enthalpy of mixing $g^M$ and the molar entropy of mixing $s^M$ are thus expressed, respectively, in Eq. 8.7:

$$g^M = RT\, x_1 \ln x_1 + RT\, x_2 \ln x_2, \qquad s^M = -\frac{\partial g^M}{\partial T} = -R\, x_1 \ln x_1 - R\, x_2 \ln x_2. \tag{8.7}$$

These simple equations for the mixing terms of the molar free enthalpy and entropy are characteristic for perfect solutions and are identical with those for ideal gas mixtures.

Further, the enthalpy of mixing of a perfect solution derived from Eq. 8.7 is zero as shown in Eq. 8.8 (vid. sections 5.9 and 5.10 ):

$$-\frac{h^M}{T} = \frac{\partial\left(\frac{g^M}{T}\right)}{\partial T} = 0, \tag{8.8}$$

indicating that the mixing of two constituents to form a perfect solution takes place at constant enthalpy, thus causing no absorption or evolution of heat at constant pressure.

The volume of mixing for a perfect solution is also zero from Eq. 8.9 (vid. sections 5.9 and 5.10):

$$v^M = \frac{\partial g^M}{\partial p} = 0, \tag{8.9}$$

indicating that the process of mixing to make a perfect solution is accompanied by neither expansion nor contraction of the solution. If we write $v_1^0$ and $v_2^0$ as the molar volume of the pure constituents 1 and 2, the mean molar volume $v$ of the solution is expressed by: $v = x_1\, v_1^0 + x_1 v_2^0$. The mean molar volume $v$ of a perfect solution is thus a linear function of molar fraction as shown in Fig. 8.1.

In contrast to a perfect solution, a solution is called an *ideal* solution, if Eq. 8.1 is valid for solute substances in the range of dilute concentrations only. Moreover, the unitary chemical potential $\mu_2^*(T,p)$ of solute substance 2 is not the same as the chemical potential $\mu_2^0(T,p)$ of solute 2 in the pure substance. $\mu_2^*(T,p) \neq \mu_2^0(T,p)$; Henry's law. For the main constituent solvent, on the other hand, the unitary chemical potential $\mu_1^*(T,p)$ is normally set to be equal to $\mu_1^0(T,p)$ in the ideal dilute solution: $\mu_1^*(T,p) = \mu_1^0(T,p)$. The free enthalpy per mole of an ideal binary solution of solvent 1 and solute 2 is thus given by Eq. 8.10:

$$\begin{aligned} g_{\text{mixture}} &= x_1\, \mu_1 + x_2\, \mu_2 = \left(x_1\, \mu_1^0 + x_2\, \mu_2^*\right) + RT\, x_1 \ln x_1 + RT\, x_2 \ln x_2 \\ &= \left(x_1\, \mu_1^0 + x_2\, \mu_2^0\right) + x_2\left(\mu_2^* - \mu_2^0\right) + RT\, x_1 \ln x_1 + RT\, x_2 \ln x_2\,, \end{aligned} \tag{8.10}$$

in which an extra term $x_2\left(\mu_2^* - \mu_2^0\right)$ for solute 2 emerges as a difference in the unitary free enthalpy of solute 2 between an ideal binary solution and a comparative perfect binary solution.

We further note that the entropy of mixing of two pure substances to form an ideal dilute solution is not equal to the so-called ideal entropy of mixing but contains an extra term $x_2\left(s_2^* - s_2^0\right)$ as shown in Eq. 8.11:

$$s^M = x_2 (s_2^* - s_2^0) - R (x_1 \ln x_1 - x_2 \ln x_2). \tag{8.11}$$

where $s_2^*$ and $s_2^0$ are the unitary partial molar entropy of solute 2 in an ideal dilute solution and the molar entropy of solute 2 in the pure substance, respectively.

Similarly, we also see a difference in the enthalpy or volume per mole between a binary perfect solution and a binary dilute ideal solution. In a perfect binary solution the enthalpy $h$ or volume $v$ per mole of the solution is identical with the mean sum in the molar enthalpy or volume of the constituent substances in their pure state, and no change in the enthalpy or volume thus occurs when we make up a perfect solution from its constituent substances: $h = x_1 h_1^0 + x_2 h_2^0$ or $v = x_1 v_1^0 + x_2 v_2^0$. In the case of an ideal dilute binary solution, on the other hand, an extra term is required to equate the enthalpy or volume per mole of the solution to the mean sum in the molar enthalpy or volume of the pure constituent substances: $h \neq x_1 h_1^0 + x_2 h_2^0$ and $v \neq x_1 v_1^0 + x_2 v_2^0$. In a dilute binary solution $h_1^* = h_1^0$ and $v_1^* = v_1^0$ for the solvent, while for the solute $h_2^0 \neq h_2^{id} = h_2^*$ and $v_2^0 \neq v_2^{id} = v_2^*$, where $h_2^{id}$ and $v_2^{id}$ are the partial molar enthalpy and volume of solute 2 and are equal to the unitary partial molar enthalpy and volume $h_2^*$ and $v_2^*$ of solute 2 in an ideal binary solution, respectively (vid. section 5.9). We hence observe an increase or decrease in the enthalpy (heat of mixing) or in the volume (expansion or contraction), when we produce a dilute binary solution from its constituent substances. This extra quantity is the enthalpy of mixing $h^M$ or the volume of mixing $v^M$ in an ideal binary solution and is given by Eq. 8.12:

$$h^M = x_2(h_2^{id} - h_2^0), \quad v^M = x_2(v_2^{id} - v_2^0), \tag{8.12}$$

where $h_2^0$ and $v_2^0$ are the molar enthalpy and volume of solute 2 in the pure substance. Note that in this section we have defined for the ideal dilute solution the unitary quantities of thermodynamic potentials with respect to the *unsymmetrical reference system* for which we refer to the following section 8.3.

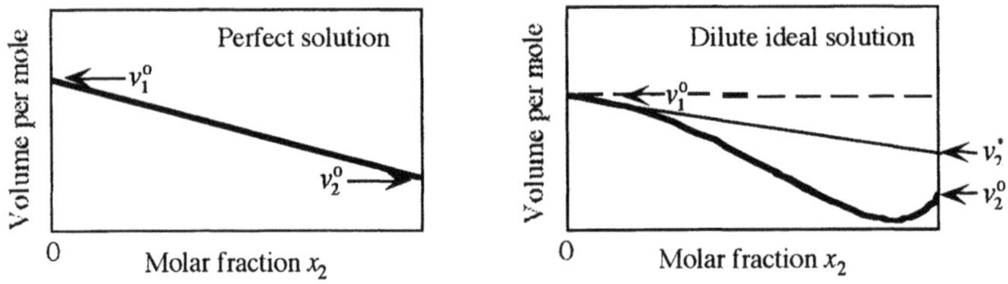


Fig. 8.1 Volume per mole as a function of the molar fraction $x_2$ of solute 2 in a binary perfect solution and in an ideal dilute solution: $v_2^*$ = the unitary partial molar volume of solute 2 extrapolated to $x_2 \rightarrow 1$.

The enthalpy and volume per mole of a binary solution both vary linearly with the molar fraction $x_2$ of solute 2 in the whole range of $x_2$ for a perfect solution and in a limited dilute range of $x_2$ for a dilute ideal solution, as schematically shown for the volume per mole of a binary solution in Fig. 8.1.

## 8. 3. Reference Systems for Thermodynamic Unitary Quantity

As mentioned in section 8.1, the value of the unitary chemical potential $\mu_i^*$ depends on the choice of the reference system. There are two reference systems which are commonly used; one is *unsymmetrical* and the other is *symmetrical*. In discussing the reference systems we shall for convenience limit ourselves to a binary solution.

We first take as a reference system an infinitely dilute solution of solute 2 in solvent 1. The chemical potentials of solvent 1 and solute 2, then, are given in the form of Eq. 8.13 for an ideal solution and in the form of Eq. 8.14 for a non-ideal solution:

$$\mu_1 = \mu_1^0 + RT \ln x_1, \quad \mu_2 = \mu_2^* + RT \ln x_2, \quad \text{ideal solution,} \tag{8.13}$$

$$\mu_1 = \mu_1^0 + RT \ln x_1 \gamma_1, \quad \mu_2 = \mu_2^* + RT \ln x_2 \gamma_2, \quad \text{non-ideal solution,} \tag{8.14}$$

where $\mu_1^0$ is the chemical potential of pure solvent 1 (equal to the unitary chemical potential $\mu_1^*$ of solvent 1 in the solution) and $\mu_2^*$ is the unitary chemical potential of solute 2 defined by the chemical potential $\mu_2^\infty$ of solute 2 extrapolated from the infinitely dilute concentration range to the unit molar fraction $x_2 = 1$; $\mu_2^* = \mu_2^\infty \neq \mu_2^0$. We have then taken the infinitely dilute solution as the reference system not only for a dilute ideal solution but also for a less dilute non-ideal solution. As the solution becomes more dilute, Eq. 8.14 approaches Eq. 8.13, which means that as $x_1 \rightarrow 1$ and $x_2 \rightarrow 0$, then $\gamma_1 \rightarrow 1$ and $\gamma_2 \rightarrow 1$. We see that these properties are *unsymmetrical*, since the two constituents 1 and 2 have not been treated in the same way.

The other choice is to define each unitary chemical potential $\mu_i^*$ as being equal to the chemical potential $\mu_i^0$ in the pure state for both solvent 1 and solute 2: $\mu_i^*(T, p) = \mu_i^0(T, p)$. We then obtain Eqs. 8.15 and 8.16 for the chemical potentials of solvent 1 and solute 2 in both an ideal and a non-ideal solution:

$$\mu_1 = \mu_1^0 + RT \ln x_1, \quad \mu_2 = \mu_2^0 + RT \ln x_2, \quad \text{ideal solution,} \tag{8.15}$$

$$\mu_1 = \mu_1^0 + RT \ln x_1 \gamma_1, \quad \mu_2 = \mu_2^0 + RT \ln x_2 \gamma_2, \quad \text{non-ideal solution,} \tag{8.16}$$

This *symmetrical* reference system gives us the activity coefficient that becomes unity as the molar fraction approaches unity for all constituent substances: $\gamma_i \rightarrow 1$ when $x_i \rightarrow 1$.

The symmetrical reference system is based on Raoult's law in a perfect solution, while the unsymmetrical reference system is based on Henry's law in an ideal dilute solution.

**8. 4. Thermodynamic Excess Functions in Non-ideal Solutions.**

For a perfect binary solution the free enthalpy (Gibbs energy) of mixing per mole has been given in Eq. 8.7. We extend this equation 8.7 to a non-ideal binary solution by using the activity coefficients $\gamma_1$ and $\gamma_2$ as shown in Eq. 8.17:

$$g^M = RT\, x_1 \ln x_1 \gamma_1 + RT\, x_2 \ln x_2 \gamma_2 . \qquad (8.17)$$

In this section we shall always define the activity coefficients with respect to the *symmetrical reference system.* Comparing Eq. 8.7 and Eq. 8.17, we define the *excess* free enthalpy (*excess* Gibbs energy) $g^E$ per mole of a non-ideal binary solution as Eq. 8.18:

$$g^E = RT\,(x_1 \ln \gamma_1 + x_2 \ln \gamma_2) . \qquad (8.18)$$

The difference in thermodynamic functions between a non-ideal solution and a comparative perfect solution is called in general the thermodynamic *excess function*. In addition to the excess free enthalpy $g^E$, other excess functions may also be defined such as excess entropy $s^E$, excess enthalpy $h^E$, excess volume $v^E$, and excess free energy $f^E$ per mole of a non-ideal binary solution. These excess functions can be derived as partial derivatives of the excess free enthalpy $g^E$ in the following.

For excess entropy $s^E$:

$$s^E = -\frac{\partial g^E}{\partial T}, \qquad s^E = -RT\left(x_1 \frac{\partial \ln \gamma_1}{\partial T} + x_2 \frac{\partial \ln \gamma_2}{\partial T}\right) - R\,(x_1 \ln \gamma_1 + x_2 \ln \gamma_2). \qquad (8.19)$$

For excess enthalpy $h^E$:

$$h^E = -T^2 \frac{\partial\left(\frac{g^E}{T}\right)}{\partial T}, \qquad h^E = -RT^2\left(x_1 \frac{\partial \ln \gamma_1}{\partial T} + x_2 \frac{\partial \ln \gamma_2}{\partial T}\right). \qquad (8.20)$$

This excess enthalpy $h^E$ corresponds to the heat of mixing of the non-ideal binary solution at constant pressure. Namely, $h^E = x_1 h_1^M + x_2 h_2^M$ with $h_i^M = h_i - h_i^0 = -RT^2(\partial \ln \gamma_i / \partial T)$, where $h_i^M$ is the partial molar heat of mixing of substance *i*, $h_i$ is the partial molar enthalpy of *i* in the non-ideal binary solution, and $h_i^0$ is the molar enthalpy of pure substance *i*. Remind ourselves that the reference system for the activity coefficients is symmetrical.

For excess volume $v^E$ we obtain Eq. 8.21:

$$v^E = \frac{\partial g^E}{\partial p}, \qquad v^E = RT\left(x_1 \frac{\partial \ln \gamma_1}{\partial p} + x_2 \frac{\partial \ln \gamma_2}{\partial p}\right). \qquad (8.21)$$

This excess volume $v^E$ is the difference between the mean molar volume of the non-ideal binary solution, $v^{nonideal} = V^{nonideal}/(n_1 + n_2)$, and the mean molar volume of the perfect binary solution $v^{perf} = x_1 v_1^0 + x_2 v_2^0$ (the sum of the volume of the two pure substances before mixing to form one mole of the solution): i.e. $v^E = v^{nonideal} - v^{perf} = v^{nonideal} - x_1 v_1^0 - x_2 v_2^0$, where $v_1^0$ and $v_2^0$ are the molar volumes of pure solvent 1 and pure solute 2, respectively.

Furthermore, for excess heat capacity $c_p^E$ at constant pressure we obtain Eq. 8.22:

$$c_p^E = \frac{\partial h^E}{\partial T}, \qquad c_p^E = -2\,RT\left(x_1\frac{\partial \ln \gamma_1}{\partial T} + x_2\frac{\partial \ln \gamma_2}{\partial T}\right) - RT^2\left(x_1\frac{\partial^2 \ln \gamma_1}{\partial T^2} + x_2\frac{\ln \gamma_2}{\partial T^2}\right). \quad (8.22)$$

This excess heat capacity $c_p^E$ is the difference between the mean molar heat capacity of the non-ideal binary solution, $c_p^{nonideal} = C_p^{nonideal}/(n_1 + n_2)$, and the mean molar heat capacity of the perfect binary solution $c_p^{perf} = x_1 c_1^0 + x_2 c_2^0$ (the sum of the heat capacities of the two pure constituent substances): i.e. $c_p^E = c_p^{nonideal} - c_p^{perf} = c_p^{nonideal} - x_1 c_{p,1}^0 - x_2 c_{p,2}^0$, where $C_p^{nonideal}$ and $c_p^{nonideal}$ are the heat capacity and the molar heat capacity of the non-ideal binary solution at constant pressure, respectively.

In the foregoing the excess function has been defined for one mole of the non-ideal solution. For the whole system of $n$ moles of substances present, we then obtain Eq. 8.23:

$$G^E = n\,g^E, \qquad H^E = n\,h^E, \qquad S^E = n\,s^E. \quad (8.23)$$

We also see that the excess free enthalpy $G^E$ is differentiated with respect to the temperature and the number of moles of the solution to give the excess entropy $S^E$ and the partial molar excess free energy of mixing $RT \ln \gamma_i$ as follows:

$$\frac{\partial G^E}{\partial T} = -S^E, \qquad \frac{\partial G^E}{\partial n_i} = RT \ln \gamma_i. \quad (8.24)$$

A discrepancy in free enthalpy between the perfect solution and the non-ideal solution, if the reference system is symmetrical, is generally expressed by the excess free enthalpy $G^E$, which consists of the enthalpy term $H^E$ and the entropy term $-TS^E$; i.e. $G^E = H^E - TS^E$. Two situations arise accordingly in non-ideal solutions depending on which of the two terms, $H^E$ and $-TS^E$, is dominant. The non-ideal solution is called ***regular***, if its deviation from the perfect solution is caused mostly by the excess enthalpy (heat of mixing) $H^E$:

$$|H^E| >> |T\,S^E|, \quad G^E \approx H^E; \quad \text{regular solutions.} \quad (8.25)$$

On the other hand, the solution is called *athermal*, if its deviation from the perfect solution is caused mostly by the excess entropy $-TS^E$ as shown in Eq. 8.26:

$$|H^E| << |T\,S^E|, \quad G^E \approx -T\,S^E; \quad \text{athermal solutions.} \quad (8.26)$$

**8. 5. Units of the Concentration.**

To express the concentration of a solution we frequently use, besides the molar fraction, the molality $m_i$, which represents the number of moles of solute $i$ in one kilogram of the solvent, and the molar concentration (molarity) $c_i$, which is the number of moles of solute $i$ per unit volume of the solution.

Of a solution containing $n_i$ moles of solute $i$ and $n_1$ moles of the solvent 1 the molality $m_i$ of solute $i$ is related to the molar fraction $x_i$ of solute $i$ as shown in Eq. 8.27:

$$m_i = \frac{1000\, n_i}{n_1 M_1} = \frac{1000\, x_i}{x_1 M_1}, \tag{8.27}$$

where $M_1$ is the molecular mass of solvent 1. For a dilute solution where $x_1 \to 1$, we have an approximate equation shown in Eq. 8.28:

$$m_i = 1000 \frac{x_i}{M_1}. \tag{8.28}$$

The chemical potential of $i$ is expressed using the molality scale as shown in Eq. 8.29:

$$\mu_i = \mu_i^{*\,m}(T, p) + RT \ln m_i \gamma_i, \tag{8.29}$$

where the unitary chemical potential of $\mu_i^{*\,m}(T,p)$ on the molality scale is related with that of $\mu_i^{*\,x}(T,p)$ on the molar fraction scale as follows:

$$\mu_i^{*\,m} = \mu_i^{*\,x}(T, p) + RT \ln \frac{M_1}{1000}. \tag{8.30}$$

Similarly, with the molar concentration scale we obtain Eqs. 8.31 and 8.32:

$$\mu_i = \mu_i^{*\,c}(T, p) + RT \ln c_i \gamma_i, \tag{8.31}$$

$$\mu_i^{*\,c}(T, p) = \mu_i^{*\,x}(T, p) + RT \ln v_1^0(T, p), \tag{8.32}$$

where $v_1^0(T,p)$ is the molar volume of the pure solvent 1 at temperature $T$ and pressure $p$. The molar concentration $c_i$ can be approximated with $x_i / v_1^0$ in dilute solutions.

### 8. 6. Osmotic Pressure.

Let us consider a semipermeable membrane separating a pure liquid solvent 1 from a solution containing solvent 1 and solute substances as shown in Fig. 8.2. The chemical potentials of solvent 1 in the pure solvent and in the solution, $\mu_1'$ and $\mu_1''$, are given by Eqs. 8.33 and 8.34, respectively:

$$\mu_1' = \mu_1^0(T, p'), \tag{8.33}$$

$$\mu_1'' = \mu_1^0(T, p'') + \phi\, RT \ln x_1, \tag{8.34}$$

where $\mu_1^0(T,p)$ is the chemical potential of the pure solvent 1 and $\phi$ is the osmotic coefficient defined in Eq. 8.4. In the *osmotic equilibrium state* the pressure $p'$ on the pure solvent side is usually higher than the pressure $p''$ on the solution side.

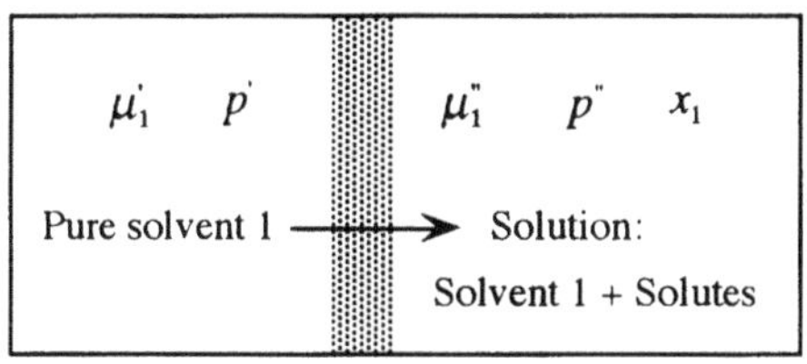


Fig. 8.2 Permeation of solvent 1 through a semipermeable membrane between a pure solvent 1 and its solution.

The driving force for osmotic permeation across the membrane is given by the affinity $A$ of the flow of solvent molecules from the pure solvent to the solution as shown in Eq. 8.35:

$$A = \mu_1' - \mu_1'' = \mu_1^0(T, p') - \mu_1^0(T, p'') - \phi\, RT \ln x_1. \tag{8.35}$$

From Eq. 7.33 we have the chemical potentials of solvent 1 in the pure solvent and in the solution as shown in Eqs. 8.36a and 8.36b, respectively:

$$\mu_1^0(T, p') = \mu_1^0(T, 0) + p'\, v_1^0 \left(1 - \frac{1}{2}\kappa p'\right), \tag{8.36a}$$

$$\mu_1^0(T, p'') = \mu_1^0(T, 0) + p''\, v_1 \left(1 - \frac{1}{2}\kappa p''\right), \tag{8.36b}$$

where $\kappa$ is the compressibility of the solution, $v_1^0$ is the molar volume of pure solvent 1, and $v_1$ is the partial molar volume of solvent 1 in the solution. Under ordinary conditions we have $v_1 \approx v_1^0$. Taking the molar volume of the pure solvent as $\overline{v_1^0} \equiv v_1^0\{1 - (1/2)\,\kappa\,(p' + p'')\}$ at the average pressure $(p' + p'')/2$, we obtain Eq. 8.37 for the *osmotic pressure* $\pi$ at osmotic equilibrium ($A = 0$ in Eq. 8.35):

$$\pi = p'' - p', \qquad \pi = -\frac{\phi\, RT \ln x_1}{\overline{v_1^0}}. \tag{8.37}$$

This equation enables the osmotic pressure to be calculated as a function of $\phi$, $x_1$, $\overline{v_1^0}$, and $T$.

In the case of an ideal solution in which $\phi = 1$, Eq. 8.37 yields Eq. 8.38:

$$\pi^{\mathrm{id}} = -\frac{RT \ln x_1}{\overline{v_1^0}}, \tag{8.38}$$

which shows that the osmotic pressure $\pi^{\mathrm{id}}$ of the ideal solution is independent of the nature of the dissolved solute. If we neglect the compressibility of the pure solvent, the average molar volume $\overline{v_1^0}$ for solvent 1 may be replaced by the usual molar volume $v_1^0$ of pure solvent 1 in Eqs. 8.37 and 8.38.

For very dilute and ideal solutions, since $\ln x_1 = \ln\left(1 - \sum x_i\right) \approx -\sum x_i$ and $c_i \approx x_1 / v_1^0$, where $i$ denotes solute substances, Eq. 8.38 with $\overline{v_1^0} = v_1^0$ yields Eq. 8.39:

$$\pi^{\mathrm{id}} = RT \sum_i c_i. \tag{8.39}$$

This is known as the *van't Hoff's law* showing that, independent of the kinds of solvents, the osmotic pressure in dilute solutions is a function of the concentration of the solutes only.

## 8. 7. Electrolytic Solutions.

Electrolytic solutions contain not only neutral molecules but also charged ions which are formed by dissociation of neutral molecules such as acetic acid shown in Eq. 8.40:

$$\mathrm{HAc} = \mathrm{H^+} + \mathrm{Ac^-}. \tag{8.40}$$

An ion charged positive is called a *cation* and an ion charged negative is called an *anion.* The total charge of cations is of course equal to the total charge of anions in any electrolytic solution which is electrically neutral as a whole.

Since an ion has an electric charge, the partial molar free enthalpy $g_i$ of an ion $i$ consists not only of the chemical potential $\mu_i$ but also of the electrostatic energy $z_i F\phi$ of the ion; where $z_i$ is the ionic valence, $F$ is the Faraday constant, and $\phi$ is the electrostatic *inner potential* of the solution. This partial molar free enthalpy $g_i$ defines the *electrochemical potential* $\eta_i$ of an ion in an electrolyte solution as shown in Eq. 8.38:

$$g_i = \eta_i = \mu_i + z_i F\phi. \tag{8.41}$$

Let us now consider the reaction of acetic acid dissociation shown in Eq. 8.37. By using the electrochemical potential of ions, the equilibrium of the reaction is expressed in Eq. 8.42:

$$\mu_{\mathrm{HAc}} = \eta_{\mathrm{H^+}} + \eta_{\mathrm{Ac^-}} = \mu_{\mathrm{H^+}} + F\phi + \mu_{\mathrm{Ac^-}} - F\phi = \mu_{\mathrm{H^+}} + \mu_{\mathrm{Ac^-}}, \tag{8.42}$$

which indicates that the sum of the electrochemical potentials of dissociated cations and anions is equal to the chemical potential of the undissociated molecules at equilibrium.

In an electrolyte solution the chemical potentials of a cation and an anion can not be measured separately. Accordingly, we define the mean chemical potential $\mu_{\pm}$ of the cation and anion by Eq. 8.43:

$$\mu_{\pm} = \frac{1}{2}\left(\mu_{H^+} + \mu_{Ac^-}\right). \tag{8.43}$$

If in an electrolyte solution a molecule dissociates into $\nu_+$ pieces of cations each with a valence $z_+$ and $\nu_-$ pieces of anions each with a valence $z_-$, the relation of $z_+\nu_+ + z_-\nu_- = 0$ holds. We can then define the mean chemical potential $\mu_{\pm}$ of the dissociated ions in general by Eq. 8.44:

$$\mu_{\pm} = \frac{\nu_+\mu_+ + \nu_-\mu_-}{\nu_+ + \nu_-}. \tag{8.44}$$

The mean chemical potential of a pair of cations and anions can be estimated from the ionic dissociation equilibrium shown as an example in Eqs. 8.42 and 8.45:

$$\mu_{HAc} = 2\,\mu_{\pm(H^+,\,Ac^-)}, \tag{8.45}$$

which enables us to estimate $\mu_{\pm(H^+,\,Ac^-)}$, if $\mu_{HAc}$ is known.

# CHAPTER 9

# ELECTROCHEMICAL ENERGY

Electrochemical energy concerns electrochemical processes in which charged particles are involved besides neutral molecules. The energy level of a charged particle is expressed by its electrochemical potential, which consists of a chemical potential and an electrostatic potential. The electrode potential frequently used in describing electrochemical energy conversion is a physical intensive variable corresponding to the energy level of electrons or ions in electrodes. This chapter discusses the energy level of charged particles, the electrode potential, the electromotive force, and the equilibrium of charge transfer reactions. Also examined is the chemical potential of hydrated ions.

## 9. 1. Electrochemical Potential of Charged Particles.

Charged particles such as ions and electrons play an important role in what is called electrochemical processes. We shall now discuss the energy level of ions and electrons in an electrochemical system. The partial molar free enthalpy (partial molar Gibbs energy) of a charged particle $i$, as described in the foregoing chapter (section 8.7), is represented by the *electrochemical potential* $\eta_i$ shown in Eq. 9.1:

$$\eta_i = \mu_i + z_i F\phi, \tag{9.1}$$

where $\mu_i$ is the chemical potential of $i$, $z_i$ is the number of elemental charge of $i$, and $\phi$ is the electrostatic *inner potential* of the electrochemical system. In physics the reference level of the electrostatic potential is usually set zero at infinite distance in vacuum.

The electrostatic inner potential $\phi$ in a condensed phase (liquid or solid) consists of the *outer potential* $\psi$ and the *surface potential* $\chi$ as shown in Fig. 9.1 and Eq. 9.2:

$$\phi = \psi + \chi. \tag{9.2}$$

The outer potential $\psi$ is an electrostatic potential at the closest distance ($\approx 1\times10^{-3}$ mm) to the surface that a charged particle can approach without being affected by any image force from the condensed phase.

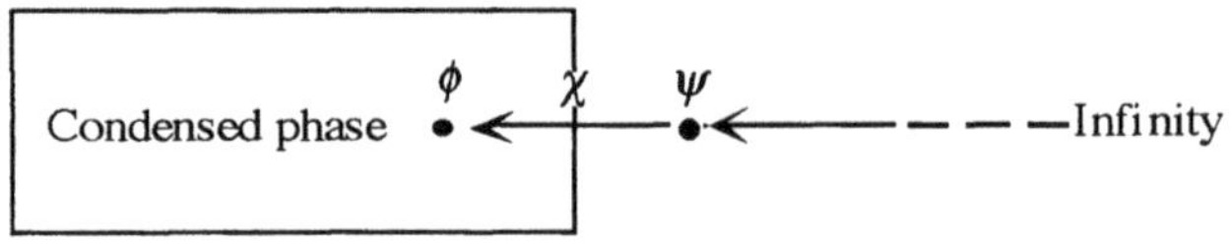


Fig. 9.1. Inner potential $\phi$, outer potential $\psi$, and surface potential $\chi$ of a condensed phase.

The electrochemical potential $\eta_i$ of a charged particle $i$ is then expressed by Eq. 9.3:

$$\eta_i = \mu_i + z_i F\phi = \mu_i + z_i F\chi + z_i F\psi = \alpha_i + z_i F\psi. \tag{9.3}$$

The outer potential $\psi$ depends on the electric charge on the condensed phase, while both the chemical potential $\mu_i$ and the surface potential $\chi$ remain constant irrespective of the electric charge. The electrochemical potential $\eta_i$ of a charged particle $i$ thus varies depending upon the amount of charge on the condensed phase, and consequently, it can not reasonably specify the energy level of the charged particle $i$ in the condensed phase.

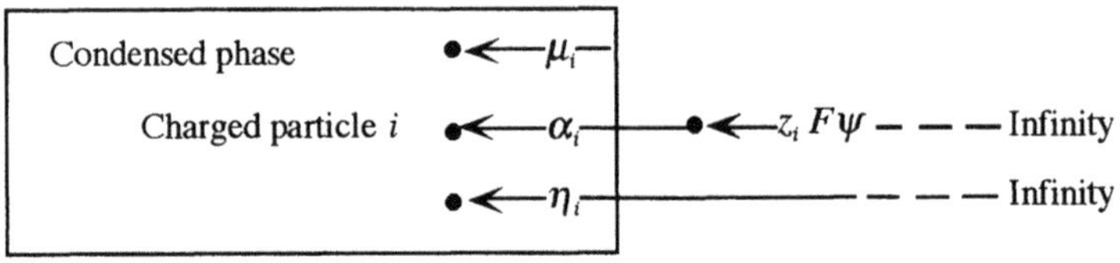


Fig. 9.2. Chemical potential $\mu_i$, real potential $\alpha_i$, and electrochemical potential $\eta_i$ of a charged particle $i$ in a condensed phase.

On the other hand, if we take as a criterion for defining the energy level of a charged particle $i$ in a condensed phase the sum of the chemical potential $\mu_i$ and the electrostatic energy of $z_i F\chi$ due to the surface potential $\chi$, this gives a uniquely defined energy level $\alpha_i$ of the charged particle $i$ in the condensed phase whatever the amount of electric charge on the phase is:

$$\alpha_i = \mu_i + z_i F\chi. \tag{9.4}$$

In electrochemistry this energy level $\alpha_i$ is called the *real potential* of charged particle *i* in a condensed phase. For electrons the real potential $\alpha_e$ is equivalent to what is called in physics the work function $\Phi$ of electrons: $-\alpha_e = \Phi$; the work function is the energy required for electron emission from a solid. Figure 9.2 shows schematically the relations between $\mu_i$, $\alpha_i$, and $\eta_i$ of a charged particle *i* in a condensed phase.

## 9. 2. Transfer of Charged Particles Between Two Condensed Phases.

If the transfer of a charged particle *i* takes place reversibly at the interface between two condensed phases 1 and 2 as shown in Fig. 9.3, the electrochemical potentials of *i* in the two phases are equilibrated to each other: $\eta_{i(1)} = \mu_{i(1)} + z_i F\phi_1 = \mu_{i(2)} + z_i F\phi_2 = \eta_{i(2)}$. A difference in the inner potential $\Delta\phi_{1/2} = \phi_1 - \phi_2$ consequently arises between the two phases as given by Eq. 9.5:

$$\Delta\phi_{1/2} = \phi_1 - \phi_2 = \frac{\mu_{i(2)} - \mu_{i(1)}}{z_i F}, \tag{9.5}$$

which may be called the *interfacial inner potential* or simply the *interfacial potential*.

Furthermore, as shown in Fig. 9.3, there arises between the two phases a difference in the outer potential $\Delta\psi_{1/2} = \psi_1 - \psi_2$, which is equivalent to what in physics called the *contact potential.* In electrochemistry we may call $\Delta\psi_{1/2}$ the *interfacial outer potential*. The relation of $\Delta\phi_{1/2}$ to $\Delta\psi_{1/2}$ is given by Eq. 9.6:

$$\Delta\phi_{1/2} = \Delta\psi_{1/2} + (\chi_1 - \chi_2), \tag{9.6}$$

where $\chi_1$ and $\chi_2$ are the surface potentials of phase 1 and phase 2, respectively. Note that the interfacial outer potential $\Delta\psi_{1/2}$ can be measured; whereas, the interfacial inner potential $\Delta\phi_{1/2}$ can not be measured by ordinary methods because of the immeasurable surface potentials not equal to each other of the two phases..

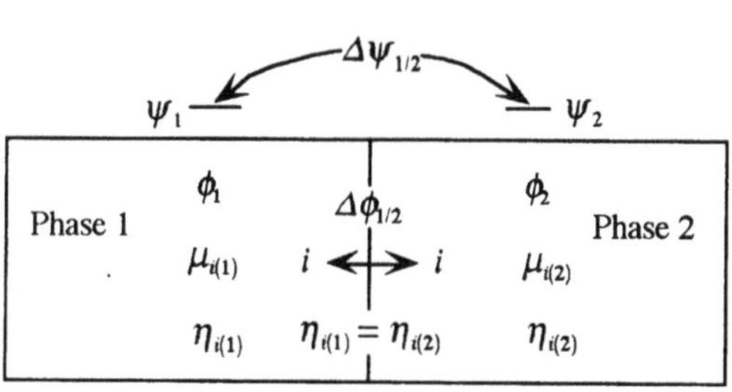


Fig. 9.3. Transfer equilibrium of charged particle *i* across an interface between a condensed phase 1 and a condensed phase 2.

**9. 3. Electrode and Electrode Potential.**

Electrochemical reactions usually occur at the interface between a solid electrode and a liquid electrolyte. The electrode is an electron conductor, such as metals and semiconductors, and is immersed in an electrolyte. In practice the electrode is partially immersed in an electrolyte, but in theory it is convenient to define that the electrode is a multiphase system in which an electronic conductor is fully immersed in an electrolyte as shown in Fig. 9.4.

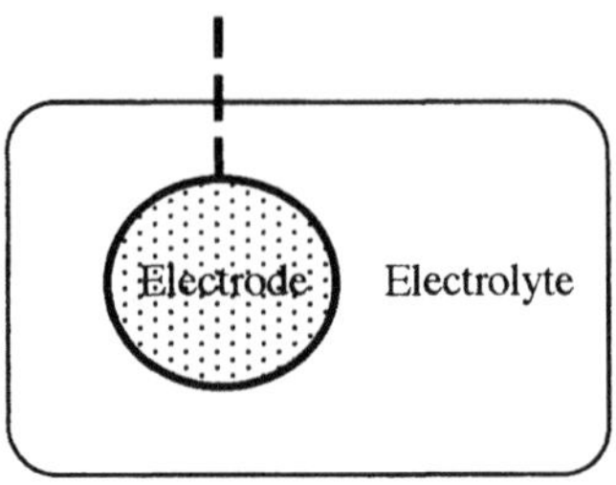


Fig. 9.4. An electrode system.

An electrode is called an *electronic electrode* when the transfer of electrons occurs, while it is called an *ionic electrode* when the transfer of ions occurs at the electrode interface. Although electrons and ions are in the same category of charged particles, they are different in electrochemical behavior due to a difference in the type of statistics that governs them. Electrons are Fermi particles which obey the Fermi statistics, whereas ions are Boltzmann particles which obey the Boltzmann statistics.

In electrochemistry we frequently refer to a technical term *electrode potential*. The electrode potential means in its physical sense the energy level, i.e. the electrochemical potential, of electrons in an electrode. It is however convenient, as described in the foregoing (Eqs. 9.3 and 9.4), to define the electrode potential in terms of the real potential $\alpha_e$ rather than the electrochemical potential $\eta_e$ of electrons in the electrode.

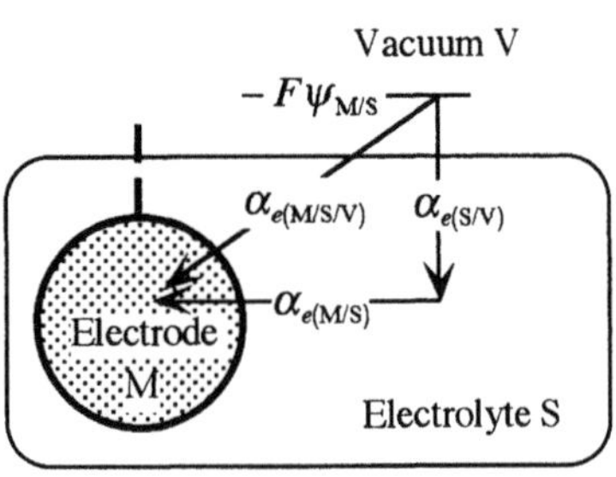


Fig. 9.5. Real potential $\alpha_{e(M/S/V)}$ of electrons in an electrode.

The *real potential* $\alpha_{e(M/S/V)}$ of an electron in an electrode is equivalent to the energy required to transfer an electron from the position of the outer potential of the electrode system to the interior of the electrode as shown in Fig. 9.5 and is given by Eq. 9.7:

$$\alpha_{e\,(M/S/V)} = \alpha_{e\,(S/V)} + \alpha_{e\,(M/S)} = \mu_{e\,(M)} - F\Delta\phi_{M/S} - F\chi_{S/V}, \tag{9.7}$$

where $\alpha_{e(S/V)}$ is the real potential of an electron in the electrolyte, $\alpha_{e(M/S)}$ is the energy required to transfer an electron from the electrolyte to the electrode, $\mu_{e(M)}$ is the chemical potential of an electron in the electrode M, $\Delta\phi_{M/S}$ is the interfacial potential of the electrode (M/S), and $\chi_{S/V}$ is the surface potential of the electrolyte (S/V). We now define the electrode potential $E$ as shown in Eq. 9.8 (Ref. 4 and 5):

$$E = \frac{\alpha_{e\,(M/S/V)}}{F} = \Delta\phi_{M/S} + \chi_{S/V} - \frac{\mu_{e\,(M)}}{F}. \tag{9.8}$$

The second and third terms on the right hands side of Eq. 9.8 remain constant for a given electrode-electrolyte system, and hence the electrode potential is a linear function of the interfacial potential $\Delta\phi_{M/S}$ of the electrode. This definition of the electrode potential holds valid for all electronic and ionic electrodes, whether the electrode reaction is in equilibrium or non-equilibrium. The potential defined by Eq. 9.8 is called the *absolute electrode potential.*

In electrochemistry we have customarily employed, instead of the absolute electrode potential $E_{\text{abs, scale}}$, a relative scale of the electrode potential, $E_{\text{hydro, scale}}$, referred to the standard or normal hydrogen electrode potential $E_{\text{NHE}}$ at which the hydrogen electrode reaction, $2H^+_{aq} + 2e_{redox} = H_{2(gas)}$, is at equilibrium in the standard state; unit activity of the hydrated proton, the standard pressure of 101.3 kPa for hydrogen gas, and room temperature of 298 K. Since $E_{\text{NHE}}$ is + 4.44 V (or + 4.5 V) in the absolute electrode potential scale, we obtain Eq. 9.9 for the relation between $E_{\text{abs, scale}}$ and $E_{\text{hydro, scale}}$ [Refs. 4 and 5.]:

$$E_{\text{hydrogen scale}} = E_{\text{absolute scale}} - 4.44\ \text{V}. \tag{9.9}$$

In the case that an electron transfer reaction (redox reaction) such as shown in Eq. 9.10 is in equilibrium at the interface of an electrode:

$$\text{RED} \rightleftarrows \text{OX} + e^-_{(\text{REDOX})}; \qquad \text{for example,} \qquad Fe^{2+} \rightleftarrows Fe^{3+} + e^-_{Fe^{3+}/Fe^{2+}}; \tag{9.10}$$

the electron $e^-_{(M)}$ in the electrode and the *redox electron* $e^-_{(REDOX)}$ of the redox particles in the electrolyte are at the same energy level so that $\alpha_{e(M/S)} = 0$, and hence $\alpha_{e(M/S/V)} = \alpha_{e(S/V)}$: no energy is required for the electron transfer between the electrode M and the electrolyte S. The electrode potential $E_{eq}$ thus corresponds to the real potential $\alpha_{e(REDOX)}$ of the redox electron in the electrolyte as shown in Eq. 9.11:

$$\alpha_{e\,(\mathrm{M/S/V})} = \alpha_{e\,(\mathrm{S/V})} = \alpha_{e(\mathrm{REDOX})} = \mu_{e(\mathrm{REDOX})} - F\,\chi_{\mathrm{S/V}}, \quad E_{\mathrm{eq}} = \frac{\alpha_{e\,(\mathrm{M/S/V})}}{-F} = \chi_{\mathrm{S/V}} - \frac{\mu_{e(\mathrm{REDOX})}}{F}, \tag{9.11}$$

where $\mu_{e(\mathrm{REDOX})}$ is the chemical potential of the redox electron in the electrolyte. The electrode potential $E_{\mathrm{eq}}$ is called the *equilibrium potential* of the redox reaction or simply the *redox potential.* It follows that the redox potential is determined by the energy level of the redox electron which is independent of the electrode material. Note that the interfacial potential, however, depends on the electrode material, since it is a function of the chemical potential of electrons in the electrode material.

In the case of an ionic electrode at which the transfer of ions, such as metal ions shown in Eq. 9.12, is in equilibrium across the electrode interface:

$$\mathrm{M}^+_{(\mathrm{M})} \rightleftarrows \mathrm{M}^+_{(\mathrm{S})}, \tag{9.12}$$

where $\mathrm{M}^+_{(\mathrm{M})}$ is the metallic ion in the metallic bonding state and $\mathrm{M}^+_{(\mathrm{S})}$ is the solvated or hydrated metallic ion in the electrolyte, the ionic transfer equilibrium determines the interfacial potential $\Delta\phi_{\mathrm{M/S}}$ of the metallic electrode, yielding $\Delta\phi_{\mathrm{M/S}}F = \mu_{\mathrm{M}^+_{(\mathrm{S})}} - \mu_{\mathrm{M}^+_{(\mathrm{M})}}$, where $\mu_{\mathrm{M}^+_{(\mathrm{S})}}$ and $\mu_{\mathrm{M}^+_{(\mathrm{M})}}$ are the chemical potentials of metal ions in the solvated state in the electrolyte and in the metallic bonding state in the electrode metal, respectively. We then obtain the real potential of electrons $\alpha_{e\,(\mathrm{M/S/V})}$ in the electrode as shown in Eq. 9.13:

$$\alpha_{e\,(\mathrm{M/S/V})} = \mu_{e\,(\mathrm{M})} - F\Delta\phi_{\mathrm{M/S}} - F\chi_{\mathrm{S/V}} = \left(\mu_{e\,(\mathrm{M})} + \mu_{\mathrm{M}^+_{(\mathrm{M})}} - \mu_{\mathrm{M}^+_{(\mathrm{S})}}\right) - F\chi_{\mathrm{S/V}},$$

$$= \left(\mu_{\mathrm{M}_{(\mathrm{M})}} - \mu_{\mathrm{M}^+_{(\mathrm{S})}}\right) - F\chi_{\mathrm{S/V}} = \mu_{e\,|\mathrm{M/M}^+|(\mathrm{S})} - F\chi_{\mathrm{S/V}} = \alpha_{e\,|\mathrm{M/M}^+|(\mathrm{S})}, \tag{9.13}$$

where $\left(\mu_{\mathrm{M}_{(\mathrm{M})}} - \mu_{\mathrm{M}^+_{(\mathrm{S})}}\right)$ is equal to the chemical potential $\mu_{e\,|\mathrm{M/M}^+|(\mathrm{S})}$ of the hypothetical electron $e^-_{|\mathrm{M/M}^+|(\mathrm{S})}$ in the electrolyte, and $\alpha_{e\,|\mathrm{M/M}^+|(\mathrm{S})}$ is the real potential of the hypothetical electron $e^-_{|\mathrm{M/M}^+|(\mathrm{S})}$ in the electrolyte; $e^-_{|\mathrm{M/M}^+|(\mathrm{S})}$ is equilibrated with the electrodic electron $e^-_{(\mathrm{M})}$ in equilibrium with the ionic reaction formally expressed by $\mathrm{M}^+_{(\mathrm{M})} + e^-_{(\mathrm{M})} \rightleftarrows \mathrm{M}_{(\mathrm{M})} \rightleftarrows \mathrm{M}^+_{(\mathrm{S})} + e^-_{|\mathrm{M/M}^+|(\mathrm{S})}$. The equilibrium potential $E_{\mathrm{eq}}$ for the metallic ion transfer is thus equivalent to the real potential $\alpha_{e\,|\mathrm{M/M}^+|(\mathrm{S})}$ of the hypothetical equilibrium electron for the metallic ion transfer as shown in Eq. 9.14:

$$E_{\mathrm{eq}} = \frac{\alpha_{e\,(\mathrm{M/S/V})}}{-F} = \frac{\alpha_{e|\mathrm{M}^+/\mathrm{M}|(\mathrm{S})}}{-F} = \chi_{\mathrm{S/V}} - \frac{\mu_{e|\mathrm{M}^+/\mathrm{M}|(\mathrm{S})}}{F}. \tag{9.14}$$

Since no electrons that pass through the electrode interface are involved in any ion transfer reactions, the hypothetical equilibrium electron for an ion transfer is virtual, and the equilibrium potential of the ion transfer reaction therefore corresponds to the energy level of that hypothetical electron in the electrolyte.

The electrode potential of an ionic electrode can also be defined by the ionic level rather than by the electronic level in the electrode. For instance, the electrode potential of a metallic electrode may be given by the real potential $\alpha_{M^{z+}(M/S/V)}$ of metallic ion in the electrode as shown in Eq. 9.15:

$$E_{ion} = \frac{\alpha_{M^{z+}(M/S/V)}}{z\,F} = \frac{\mu_{M^{z+}(M)}}{z\,F} + \Delta\phi_{M/S} + \chi_{S/V}, \qquad (9.15)$$

where $\mu_{M^{z+}(M)}$ is the chemical potential of metallic ion in the electrode metal and $z$ is the ionic valency of the metal ion. If the transfer of metal ion is in equilibrium at the electrode interface, $\alpha_{M^{z+}(M/S/V)}$ is equilibrated with the real potential $\alpha_{M^{z+}(/S/V)}$ of solvated metal ion in the electrolyte, and hence the equilibrium potential of the metallic ion transfer is equivalent to the energy level of the solvated or hydrated metallic ion in the electrolyte.

We may call the electrode potential defined by the ionic energy level the *ionic electrode potential*, and the electrode potential defined by the electronic energy level may be called the *electronic electrode potential.* In the case in which the electrode has no electronic level in the energy range of our interest such as certain membrane electrodes, it is convenient to describe the system in terms of the ionic electrode potential rather than the electronic electrode potential [Refs. 4 and 5.].

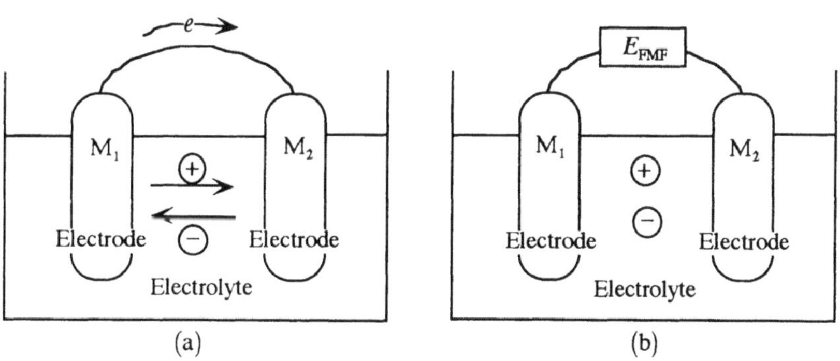


Fig. 9.6. Electrochemical cell: (a) non-equilibrium cell, (b) equilibrium cell; $E_{EMF}$= electromotive force.

## 9. 4. Electrochemical Cells.

Two electrodes, if connected, constitute an electrochemical cell as shown in Fig. 9.6. When the electrode potentials of the two electrodes differ from each other, a current flows through the cell with each electrode undergoing an electrochemical reaction, i.e. the transfer of electrons or ions. An electrode is called an *anode* at which the electrochemical reaction carries positively charged particles from the electrode to the electrolyte and negatively charged particles in the reverse direction; while on the other hand an electrode is called a *cathode* if it carries negatively charged particles from the electrode to the electrolyte and positively charged particles in the reverse direction.

The anodic reaction is an oxidation reaction producing electrons in the anode, while the cathodic reaction is a reduction reaction consuming electrodic electrons at the cathode interface. We shall consider, as an example, an electrochemical cell consisting of a metallic zinc electrode and a metallic copper electrode, in which the anodic reaction of zinc ion transfer (zinc dissolution) is coupled with the cathodic reaction of copper ion transfer (copper deposition) as shown in the following processes:

$$Zn \rightarrow Zn^{2+}_{aq} + 2\,e^{-}, \quad \text{anodic reaction (oxidation reaction).}$$

$$Cu^{2+}_{aq} + 2\,e^{-} \rightarrow Cu, \quad \text{cathodic reaction (reduction reaction).}$$

These reactions compose a whole cell reaction given as follows:

$$Zn + Cu^{2+}_{aq} \rightarrow Cu + Zn^{2+}_{aq}.$$

In order to make the cell current zero we need to put an electrostatic voltage $E_{EMF}$ in the cell circuit; Fig. 9.6. This electrostatic voltage $E_{EMF}$ is called the *electromotive force* of the cell.

The electrochemical cell is often described by a cell diagram such as shown in Eq. 9.16:

$$Zn \mid Zn^{2+} \ \vdots\ Cu^{2+} \mid Cu, \qquad (9.16)$$

where two vertical lines indicate the electrode interfaces and a vertical dotted line shows the contact of two electrolytes. From left to right this diagram is built up such that positively charged particles are transported from the electrode (anode) on the left hand side through the electrolyte to the electrode (cathode) on the right hand side, according to IUPAC recommendation. Furthermore, the electromotive force takes its reference level at the electrode potential of the electrode on the left hand side .

Let us consider an electrochemical cell shown in Eq. 9.17:

$$Pt \mid H_{2\,(gas)} \mid H_2O,\ H^{+}_{(aq)} : H_2O,\ OH^{-}_{(aq)} \mid O_{2\,(gas)} \mid Pt, \qquad (9.17)$$

whose overall cell reaction is given by Eq. 9.18:

$$H_{2\,(gas)} + \frac{1}{2}\,O_{2\,(gas)} = H_2O_{\,(aq)}. \qquad (9.18)$$

The anodic reaction on the left electrode and the cathodic reaction on the right electrode are then expressed by Eqs. 9.19 and 9.20, respectively:

$$H_{2\,(gas)} = 2\,H^{+}_{(aq)} + 2\,e^{-}_{(L)}, \qquad \text{left electrode.} \tag{9.19}$$

$$\frac{1}{2}\,O_{2\,(gas)} + 2\,H^{+}_{(aq)} + 2\,e^{-}_{(R)} = H_2O_{\,(aq)}, \qquad \text{right electrode.} \tag{9.20}$$

If the electrochemical reactions at the two electrodes are both in equilibrium, the electrochemical potentials of electrons in the two electrodes are given by Eqs. 9.21 and 9.22, respectively:

$$\eta_{e\,(L)} = \frac{1}{2}\left(\mu_{H_2\,(gas)} - 2\,\eta_{H^{+}_{(aq)}}\right), \tag{9.21}$$

$$\eta_{e\,(R)} = \frac{1}{2}\left(\mu_{H_2O\,(aq)} - 2\,\eta_{H^{+}_{(aq)}} - \frac{1}{2}\,\mu_{O_2\,(gas)}\right). \tag{9.22}$$

In any electrochemical cell the outer potentials of the electrolyte for the two electrodes are identical so long as the two electrodes are immersed in a homogeneous electrolyte. The difference in the electrochemical potential of electrons consequently becomes equal to the difference in the real potential of electrons between the two electrodes: that is $(\eta_{e(R)} - \eta_{e(L)}) = (\alpha_{e(R)} - \alpha_{e(L)})$. The electromotive force $E_{H_2/O_2}$ of the electrochemical cell is thus given by Eq. 9.23:

$$E_{H_2/O_2} = \frac{1}{-F}\left(\alpha_{e\,(R)} - \alpha_{e\,(L)}\right) = \frac{1}{-2\,F}\left(\mu_{H_2O\,(aq)} - \frac{1}{2}\,\mu_{O_2\,(gas)} - \mu_{H_2\,(gas)}\right). \tag{9.23}$$

The parenthesis on the right hand side of Eq. 9.23 is equal to minus the affinity $A_{H_2/O_2}$ of the overall cell reaction 9.18 as shown in Eq. 9.24:

$$\left(\mu_{H_2O\,(aq)} - \frac{1}{2}\,\mu_{O_2\,(gas)} - \mu_{H_2\,(gas)}\right) = -\,A_{H_2/O_2}. \tag{9.24}$$

The electromotive force $E_{H_2/O_2}$ is thus related to the reaction affinity:

$$E_{H_2/O_2} = \frac{A_{H_2/O_2}}{2\,F}. \tag{9.25}$$

The affinity of the reaction, as has been shown in the foregoing chapters 5 and 6, consists of the unitary affinity $A^{*}_{H_2/O_2}$ and the affinity of mixing $A^{M}_{H_2/O_2} = RT\ln\left(p_{H_2}\,p^{1/2}_{O_2}\right)$. We see then that the electromotive force $E_{H_2/O_2}$ also consists of the unitary electromotive force $E^{*}_{H_2/O_2}$ and the mixing term $(1/2)\,(RT/F)\ln\left(p_{H_2}\,p^{1/2}_{O_2}\right)$ as shown in Eq. 9.26:

$$E_{H_2/O_2} = E^{*}_{H_2/O_2} + \frac{RT}{2\,F}\ln\left(p_{H_2}\,p^{1/2}_{O_2}\right), \tag{9.26}$$

where $E^{*}_{H_2/O_2}$ is the unitary term at the standard pressure for both hydrogen and oxygen gas. The unitary electromotive force in the standard state is usually called the standard electromotive force, $E^{0}_{H_2/O_2}$, whose value is given by $E^{0}_{H_2/O_2} = 1.23$ V.

In general, the electromotive force $E_{EMF}$ of an electrochemical cell is given by Eq. 9.27:

$$E_{EMF} = \frac{A_{cell}}{n\,F} = E^{*}_{EMF} + \frac{1}{n\,F} RT \sum_{i} \nu_i \ln a_i \,, \tag{9.27}$$

where $A_{cell}$ is the affinity of the cell reaction, $n$ is the number of elemental charges involved in the reaction, $\nu_i$ is the stoichiometrical coefficient of particle $i$, and $a_i$ is the activity of particle $i$ taking part in the reaction.

The temperature dependence of the reaction affinity is given by Eq. 4.8, and the result for the hydrogen-oxygen cell is shown in Eq. 9.28:

$$A_{H_2/O_2} = -\,\Delta H_{H_2/O_2} + T\left(\frac{\partial A_{H_2/O_2}}{\partial T}\right)_p . \tag{9.28}$$

We then obtain from Eqs. 9.25 and 9. 28 the relation between the electromotive force $E_{H_2/O_2}$ and the enthalpy change $\Delta H_{H_2/O_2}$ of the cell reaction as shown in Eq. 9.29:

$$\Delta H_{H_2/O_2} = 2\,F\,T\left(\frac{\partial E_{H_2/O_2}}{\partial T}\right)_p - 2\,F\,E_{H_2/O_2}. \tag{9.29}$$

This equation indicates that, if the electromotive force has a positive sign and the temperature coefficient of the electromotive force of the reaction has a negative sign, the reaction enthalpy will be negative: $\Delta H_{H_2/O_2} < 0$ and hence the reaction is exothermic.

The reaction of the hydrogen-oxygen fuel cell shown in Eq. 9.18 is exothermic with the reaction enthalpy equal to $\Delta H^0_{H_2/O_2} = -47.3\ \mathrm{kJ \cdot mol^{-1}}$ at the standard temperature and pressure. The heat of reaction for gaseous hydrogen oxidation at room temperature amounts to $-284.7\ \mathrm{kJ \cdot mol^{-1}}$, in which $237.3\ \mathrm{kJ \cdot mol^{-1}}$ is used for producing the electromotive force of 1.23 V and $47.3\ \mathrm{kJ \cdot mol^{-1}}$ is exhausted as heat. On the other hand, the reaction of water electrolysis, which is the reverse reaction of the hydrogen-oxygen fuel cell reaction, absorbs an amount of energy of $237.3\ \mathrm{kJ \cdot mol^{-1}}$ equivalent to the electromotive force of 1.23 V from an electric source and an amount of $47.3\ \mathrm{kJ \cdot mol^{-1}}$ from the environment to obtain the total amount of energy $284.7\ \mathrm{kJ \cdot mol^{-1}}$ required for the dissociation of one mole of water. During water electrolysis both enthalpy and entropy increase, in which the increase in the entropy is provided for by heat absorption from the environment.

## 9. 5. Equilibrium Electrode Potential of Electronic Transfer Reactions.

Let us consider an electronic transfer reaction of the redox couple of ferrous-ferric ions:

$$Fe^{3+}_{(aq))} + e^- = Fe^{2+}_{(aq))}. \tag{9.30}$$

Connecting this redox reaction in a half cell (an electrode) on the right hand side with the standard hydrogen electrode reaction $H^+_{aq} + e^- = 0.5\,H_{2(gas)}$ in another half cell on the left side, we make up the cell shown in Fig. 9.7 and represent it in the cell diagram of Eq. 9.31:

$$\mathrm{Pt}\,|\,H_{2(gas)}\,|\,H_2O, H^+_{(aq)}\ \vdots\ Fe^{3+}_{(aq)}, Fe^{2+}_{(aq)}\,|\,\mathrm{Pt}, \tag{9.31}$$

whose total cell reaction is given by Eq. 9.32:

$$\frac{1}{2}H_{2\,(gas)} + Fe^{3+}_{(aq)} = H^+_{(aq)} + Fe^{2+}_{(aq)}. \tag{9.32}$$

The electromotive force $E_{EMF}$ of this cell is equivalent to the *redox potential* $E_{Fe^{3+}/Fe^{2+}}$ of the reaction referred to the standard hydrogen electrode, i.e. the equilibrium potential of the redox reaction, and it is given by Eq. 9.33 at the standard temperature and pressure:

$$E_{Fe^{3+}/Fe^{2+}} = \frac{A_{Fe^{3+}/Fe^{2+}}}{F} = \frac{1}{F}\left(\mu_{Fe^{3+}(aq)} + \frac{1}{2}\mu^0_{H_2\,(gas)} - \mu_{Fe^{2+}(aq)} - \mu^0_{H^+(aq)}\right)$$

$$= \frac{1}{F}\left(\mu^0_{Fe^{3+}(aq)} + \frac{1}{2}\mu^0_{H_2\,(gas)} - \mu^0_{Fe^{2+}(aq)} - \mu^0_{H^+(aq)}\right) + \frac{RT}{F}\ln\left(\frac{a_{Fe^{3+}}}{a_{Fe^{2+}}}\right)$$

$$= E^0_{Fe^{3+}/Fe^{2+}} + \frac{RT}{F}\ln\left(\frac{a_{Fe^{3+}}}{a_{Fe^{2+}}}\right), \tag{9.33}$$

where $E^0_{Fe^{3+}/Fe^{2+}}$ is the unitary redox potential at the standard state and is called the *standard redox potential* of the reaction.

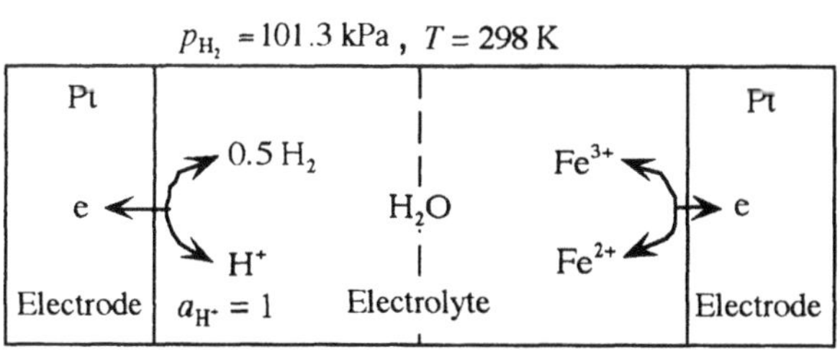


Fig. 9.7. An electrochemical cell consisting of a redox electrode reaction of hydrated ferric-ferrous ions and the standard hydrogen electrode reaction.

Table 9.1 shows the numerical values of the standard redox potentials for a few reactions of electronic transfer at electrodes. Electrochemical handbooks provide the standard redox potentials for various other transfer reactions of redox electrons. As mentioned in section 9.3, the redox potential is independent of the electrode materials.

Table 9.1. Standard redox potentials $E^0_{redox}$ of electronic electrode reactions: $E^0_{redox}$ is referred to the standard hydrogen electrode potential and $e^-_{redox}$ is the redox electron in equilibrium.

| Redox reaction | Standard redox potential $E^0_{redox}$/ $V_H$ |
|---|---|
| $2\,H^+_{aq} + 2\,e^-_{redox} = H_{2(gas)}$ | 0.0 |
| $Cu^{2+}_{aq} + e^-_{redox} = Cu^+_{aq}$ | +0.153 |
| $Fe^{3+}_{aq} + e^-_{redox} = Fe^{2+}_{aq}$ | +0.771 |
| $O_{2(gas)} + 4\,H^+_{aq} + 4\,e^-_{redox} = 2\,H_2O_{aq}$ | +1.229 |
| $Ce^{4+}_{aq} + e^-_{redox} = Ce^{3+}_{aq}$ | +1.74 |

**9. 6. Equilibrium Electrode Potential of Ionic Transfer Reactions.**

Let us now consider a transfer reaction of iron ions $Fe^{2+}_{(M)}$ from the lattice site in a metallic iron electrode to the hydrated state of iron ion $Fe^{2+}_{(aq)}$ in an aqueous solution at the standard temperature 298 K and pressure 101.3 kPa as shown in Eq. 9.34:

$$Fe^{2+}_{(aq)} = Fe^{2+}_{(M)}. \qquad (9.34)$$

Connecting this ionic transfer reaction with the standard hydrogen electrode reaction $H^+_{(aq)} + e^-_{(M)} = (1/2)H_{2(gas)}$ gives the cell shown in Fig. 9.8 and in Eq. 3.35:

$$Pt\,|\,H_{2(gas)}\,|\,H_2O, H^+_{(aq)}\ \vdots\ Fe^{2+}_{(aq)}\,|\,Fe\,. \qquad (9.35)$$

The total reaction in this cell is given by Eq. 9.36:

$$H_{2\,(gas)} + Fe^{2+}_{(aq)} = 2\,H^+_{(aq)} + \left(Fe^{2+}_{(M)} + 2\,e^-_{(M)}\right). \qquad (9.36)$$

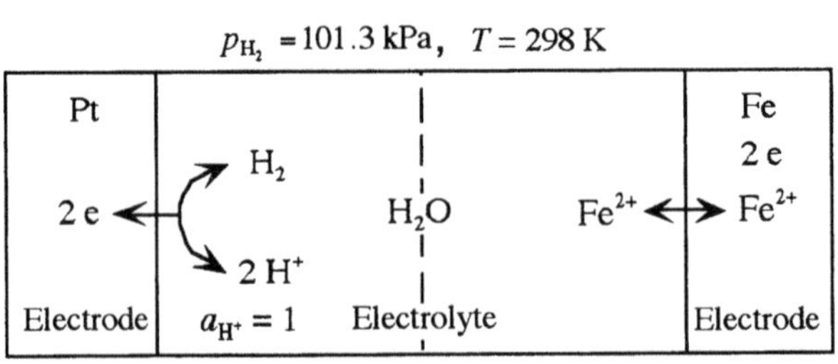


Fig. 9.8. An electrochemical cell consisting of an electrode reaction of iron ion transfer and the hydrogen electrode reaction both in equilibrium.

The electromotive force $E_{EMF}$ of this cell is equivalent to the equilibrium potential $E_{Fe^{2+}/Fe}$ of the transfer reaction of divalent iron ions referred to the standard hydrogen electrode, and it is given by Eq. 9.37 at the standard temperature and pressure :

$$E_{Fe^{2+}/Fe} = \frac{A_{Fe^{2+}/Fe}}{2F} = \frac{1}{2F}\left(\mu_{Fe^{2+}_{(aq)}} + \mu^0_{H_2\,(gas)} - \mu^0_{Fe(M)} - 2\,\mu^0_{H^+_{(aq)}}\right)$$

$$\frac{1}{2F}\left(\mu^0_{Fe^{2+}_{(aq)}} + \mu^0_{H_2\,(gas)} - \mu^0_{Fe(M)} - 2\,\mu^0_{H^+_{(aq)}}\right) + \frac{RT}{2F}\ln a_{Fe^{2+}_{(aq)}}$$

$$= E^0_{Fe^{2+}/Fe} + \frac{RT}{2F}\ln a_{Fe^{2+}_{(aq)}}, \qquad (9.37)$$

where $E^0_{Fe^{2+}/Fe}$ is the *standard equilibrium potential* of the iron ion transfer reaction at metallic iron electrodes, i.e. metallic iron dissolution and deposition.

Table 9.2 shows the numerical values of the standard equilibrium potentials for a few reactions of ion transfer at ionic electrodes. Electrochemical handbooks provide us with the standard equilibrium potential for a number of ionic transfer reactions.

Table 9.2. Standard equilibrium potential $E^0_{ion}$ of ionic electrode reactions: $E^0_{ion}$ is referred to the standard hydrogen electrode potential and $e^-_{ion}$ is the equilibrium electron in the ion transfer reactions.

| Ionic transfer reaction | Standard equilibrium potential $E^0_{ion}$/ $V_H$ |
|---|---|
| $Fe^{2+}_{aq} + 2\,e^-_{ion} = Fe_{(solid)}$ | $-0.44$ |
| $NiO_{(solid)} + 2\,H^+_{aq} + 2\,e^-_{ion} = Ni_{(solid)} + H_2O_{aq}$ | $+0.132$ |
| $AgCl_{(solid)} + e^-_{ion} = Ag_{(solid)} + Cl^-_{aq}$ | $+0.132$ |
| $Cu^{2+}_{aq} + 2\,e_{ion} = Cu_{(solid)}$ | $+0.337$ |
| $Au^{3+}_{aq} + 3\,e^-_{ion} = Au_{(solid)}$ | $+1.50$ |

## 9. 7. Chemical Potential of Hydrated Ions.

Equation 9.37 gives us the chemical potential $\mu^0_{Fe^{2+}_{(aq)}}$ of hydrated ferrous ion $Fe^{2+}_{(aq)}$ in the standard state as a function of the standard equilibrium potential $E^0_{Fe^{2+}/Fe}$ of the dissolution-deposition reaction of metallic iron as shown in Eq. 9.38:

$$\mu^0_{Fe^{2+}_{(aq)}} = 2\,F\,E^0_{Fe^{2+}/Fe} - \mu^0_{H_2\,(gas)} + \mu^0_{Fe(M)} + 2\,\mu^0_{H^+_{(aq)}}. \qquad (9.38)$$

In chemical thermodynamics the standard chemical potentials $\mu^0_{H_2\,(gas)}$ and $\mu^0_{Fe(M)}$ of hydrogen molecules and metallic iron are set zero, so that if the standard chemical potential $\mu^0_{H^+_{(aq)}}$ of hydrogen ions is known, we can estimate the standard chemical potential $\mu^0_{Fe^{2+}_{(aq)}}$ of the hydrated ferrous ion from the standard equilibrium potential $E^0_{Fe^{2+}/Fe}$.

In electrochemistry we make it a rule that the standard chemical potential $\mu^0_{H^+_{(aq)}}$ of hydrogen ions is set zero as the level of reference for the chemical potentials of all other hydrated ions. The standard chemical potentials of various hydrated ions tabulated in electrochemical handbooks are thus relative to the standard chemical potential of hydrogen ions at unit activity in aqueous solutions. Table 9.3 shows the numerical values of the standard chemical potential $\mu^0_i$, the standard partial molar enthalpy $h^0_i$, and the standard partial molar entropy $s^0_i$ for a few of hydrated ions.

Table 9.3. Standard chemical potential $\mu^0_i$, standard partial molar enthalpy $h^0_i$, and standard partial molar entropy $s^0_i$ for a few hydrated ions: Standard state; 101.3 kPa, 298 K, unit activity in molality scale.

| Ion | State | $\mu^0_i/\mathrm{kJ\cdot mol^{-1}}$ | $h^0_i/\mathrm{kJ\cdot mol^{-1}}$ | $s^0_i/\mathrm{J\cdot K^{-1}\cdot mol^{-1}}$ |
|---|---|---|---|---|
| $CO_3^{2-}$ | hydrated ion | - 527.9 | - 677.1 | - 56.9 |
| $Cl^-$ | hydrated ion | - 131.2 | - 167.2 | 56.5 |
| $Fe^{2+}$ | hydrated ion | - 78.9 | - 89.1 | - 137.7 |
| $Fe^{3+}$ | hydrated ion | - 4.7 | - 48.5 | - 137.7 |
| $H^+$ | hydrated ion | 0.0 | 0.0 | 0.0 |
| $Na^+$ | hydrated ion | - 261.9 | - 240.1 | 59.0 |
| $OH^-$ | hydrated ion | - 157.3 | - 230.0 | -10.7 |
| $SO_4^{2-}$ | hydrated ion | - 744.6 | - 909.3 | 20.1 |
| $HS^-$ | hydrated ion | 12.06 | - 17.6 | 63.8 |
| $Zn^{2+}$ | hydrated ion | - 147.1 | - 153.9 | - 112.1 |

# CHAPTER 10

# EXERGY

The law of conservation of energy indicates that energy never disappears, while the second law of thermodynamics sets forth that thermal energy can not be fully utilized so far as we are in our atmospheric environment. Engineering thermodynamics has recently introduced a new energy quantity called *exergy* to figure out how much work or power we can utilize from a given amount of energy with respect to the natural environment. Energy is conserved in any processes; whereas, exergy is dissipated in spontaneous processes. Exergy analyses are thus effective in improving the energy efficiency in practical manufacturing processes. This chapter examines the concept and property of exergy in thermodynamic systems and shows that exergy losses are in principle related to the affinity (the free enthalpy change) of irreversible processes. Also discussed are various examples of exergy evaluation for heat transfer, gas expansion, substance mixing, and chemical reactions.

### 10. 1. The Concept of Exergy.

The first law of thermodynamics has provided the concept of internal energy and enthalpy which are conserved in a physicochemical system, and the second law has defined entropy, free energy (Helmholtz energy) and free enthalpy (Gibbs energy) as thermodynamic energy functions to indicate the possibility and capacity of advancing irreversible processes. Recently (1950's), engineering thermodynamics has introduced, from a practical point of view, a new energy function called "*exergy*" to express; the amount of available energy; its ability to be converted into other kinds of energy; and especially the capacity for doing work that we can utilize with a given system of energy carriers in our normal environment on the earth [Refs. 6 and 7.]. The term of *exergy* was first introduced by Rant [Ref. 8.]; it means the amount of work (-erg) that is released (ex- ).

Exergy is defined by Szargut [Ref. 9.] as the maximum amount of work obtainable when an energy carrier is brought from its initial state to a state of thermodynamic equilibrium (an inert state) with the common substances of the natural environment by means of reversible processes, involving interaction only with the above-mentioned substances of nature. In other words, exergy is the maximum amount of work that we can obtain if a substance or a form of energy is converted to its inert reference state. Exergy can thus be also the minimum amount of work to be supplied if a substance or a form of energy has to be produced from its inert reference state. Exergy is expressed in terms of state functions alone as far as it is defined by means of reversible processes, and then it is a state function. Furthermore, as will be shown later, exergy bears a close resemblance to the energy function of *affinity* (free enthalpy change) which, as mentioned in chapter 4, represents the maximum work (available energy) that we can draw out from a physical or chemical process of energy carriers when advancing reversibly from its initial state toward its final equilibrium state. In contrast to affinity, however, the exergy of energy carriers fixes its final reference level at the state of the most stable existence (the dead state, the inert reference state) that the energy carriers can reach in the ordinary natural environment on the earth [Refs. 7 and 9.]; the dead state or the inert reference state is the state of chemical elements in which they are in "unlimited supply" without expenditure of work from our natural environment that is in equilibrium with the exergy reference substances. The exergy of any material is then the minimum expenditure of work required to obtain this material from the reference level substances in our natural environment [Ref. 10.].

Exergy therefore results from a difference in free enthalpy (Gibbs energy) between the energy carriers under consideration and the common reference substances in the natural environment: exergy is thus a function of the thermodynamic state of the substances under consideration and of the thermodynamic state of the common reference substances in the natural environment. In other words, exergy arises from an interaction between the substances under consideration and the common reference substances in the environment.

Seawater on the earth is well known to possess a huge amount of internal energy or enthalpy, which we can not utilize on the earth in global equilibrium with the ocean, however. In other words, the affinity of seawater toward the global earth environment is zero, and seawater thus possesses no exergy. On the other hand, substances at a temperature higher or lower than the atmospheric temperature on the earth contain an amount of available energy for work, and they hence have an exergy value greater than zero. Analogously, gases at a pressure higher or lower than the atmospheric pressure also possess an amount of exergy.

Gaseous molecular oxygen in the atmospheric air holds itself at the molar fraction of $x_{O_2} = 0.21$, and it is at this concentration that the exergy of gaseous oxygen is zero. In order to produce pure oxygen gas from air, then, an amount of free enthalpy is required equivalent to minus the chemical potential of mixing for oxygen molecules in air, so that the pure oxygen gas possesses an amount of exergy greater than zero. Similarly, pure metallic iron possesses an amount of exergy equivalent to the affinity of the chemical reaction to form iron oxide from metallic iron and oxygen gas in air, since metallic iron spontaneously corrodes changing into iron oxide that is the most stable existence of iron in the natural environment (a top layer of the lithosphere).

Any spontaneous change of substances that occurs in the natural environment advances with a decrease in exergy of the substances: this is the law of exergy decrease in spontaneous processes in analogy to the law of affinity decrease in spontaneous processes. In contrast to energy which is always conserved in any processes due to the first law of thermodynamics, exergy is exempt from the law of conservation and so is the affinity.

The exergy of a substance is conventionally classified into two parts: *physical exergy* associated with changes in temperature (thermal exergy), changes in pressure (pressure exergy, dynamic exergy), and changes in concentration (mixing exergy); and *chemical exergy* associated with changes in chemical composition of substances.

In this book we shall use the symbol epsilon, $E$ and $\varepsilon$, in Greek to express total exergy and molar exergy (or specific exergy), respectively.

## 10. 2. Exergy and Heat.

Thermal energy (heat), as has been mentioned in the foregoing (e.g. section 3.7), can be converted only partly into work, whereas the other forms of energy are theoretically all convertible into one another. We now consider the amount of work generated from an amount of thermal energy at a high temperature level $T$ with respect to our environment at temperature $T_0$. An ideal reversible heat engine (section 3.7) converts thermal energy into work as shown in Fig. 10.1, generating the maximum amount of work, $W_{rev}$, that can be obtained from an amount of thermal energy $Q$ received by a working substance (an ideal gas) in the engine from an outside heat source at a high temperature $T$ and releasing an amount of thermal energy $Q_0$ into an outside heat reservoir at a low temperature $T_0$ in the absolute temperature scale:

$$W_{rev} = Q - Q_0 = Q\,\frac{T - T_0}{T}. \qquad (10.1)$$

If $T_0$ is set at the temperature of our natural environment to which exergy is referred, $W_{rev}$ becomes equivalent to the exergy of the thermal energy at temperature $T$. The exergy, $E$, of an amount of heat $Q$ at a high temperature $T$ is thus evaluated at Eq. 10.2:

$$E = W_{rev} = Q\,\frac{T - T_0}{T}, \qquad (10.2)$$

which is the maximum amount of work that we can gain reversibly from an amount of thermal energy, $Q$, at temperature $T$.

The thermal energy, $Q_0 = Q - E = Q(T_0/T)$, released from the engine to the environment at temperature $T_0$ can not be used and is called ***anergy*** in terms of engineering thermodynamics. In general, anergy is defined as the difference in amount between energy and exergy:

$$\text{Anergy} = \text{Energy} - \text{Exergy}. \qquad (10.3)$$

The efficiency of the reversible heat engine, $\lambda_{rev} = (W_{rev}/Q)$ as given in Eq. 3.45, represents the *energy availability* $\lambda_Q$ of an amount of heat $Q$ at a constant temperature $T$:

$$\lambda_Q = \frac{W_{rev}}{Q} = \frac{E}{Q} = 1 - \frac{T_0}{T}. \tag{10.4}$$

When the heat engine is operated not in a reversible but in an irreversible way, the amount of work, $W_{irr}$, that we obtain from the amount of thermal energy $Q$ is less than the maximum amount of work, $W_{rev}$, and hence it is less than the exergy, $E$, of the thermal energy $Q$ at temperature $T$ as shown in Eq. 105:

$$W_{irr} = E - \Delta E_{irr} < W_{rev}, \tag{10.5}$$

where $\Delta E_{irr}$ is the amount of exergy that has been lost in the irreversible heat engine and is called *exergy loss* due to the irreversibility in the energy transformation. In other words, to obtain the same amount of work we need a greater amount of driving exergy in irreversible processes than in the reversible process.

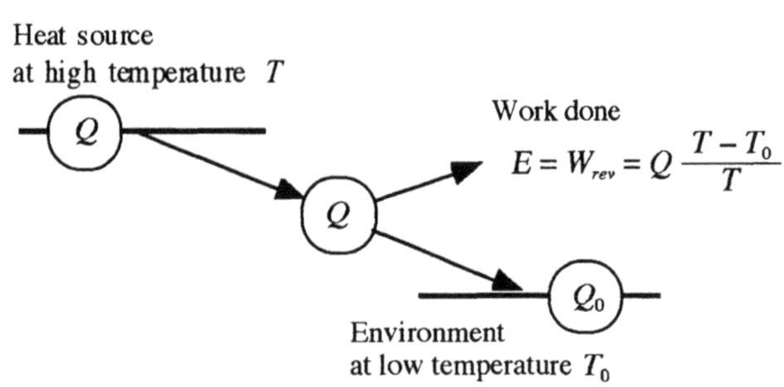


Fig. 10.1. Conversion of heat $Q$ into work $W_{rev}$ through a reversible heat engine between a high temperature $T$ and the temperature $T_0$ of our environment.

## 10. 3 Exergy and Pressure.

Let us now consider a gas phase of volume $V$ and pressure $p$ that expands itself reversibly at constant temperature $T_0$ toward the state of volume $V_0$ at pressure $p_0$ in equilibrium with the atmosphere. The reversible work done by the gas is then given by Eq. 10.6:

$$W_{rev} = \int_{p}^{p_0} p\, dV. \tag{10.6}$$

Since the work done against the atmospheric pressure $p_0$ of the outside environment, i.e. the work done for removing a volume of atmospheric gas, can not be utilized, the available work equivalent to exergy $E$ will be less than $W_{rev}$ as shown in Fig. 10.2. If $p$ is much greater than $p_0$, however, we may assume that $W_{rev}$ approximately equals $E$. From the equation of state for an ideal gas, $pV = nRT_0$, with $n$ being the number of moles of the gas, we obtain $pdV + Vdp = 0$ and $dV = -(V/p)\,dp = -(n\,R\,T_0/p^2)\,dp$. Substituting these equations into Eq. 10.6, then, yields Eq. 10.7 for the exergy of an ideal gas expressed as a function of its pressure:

$$E = W_{rev} = -n\,R\,T_0 \int_p^{p_0} \frac{dp}{p} = n\,R\,T_0 \ln \frac{p}{p_0}. \tag{10.7}$$

From Eqs. 10.7 and 3.47 the pressure-dependent molar exergy $\varepsilon$ of an ideal gas is:

$$\varepsilon = R\,T_0 \ln \frac{p}{p_0} = -T_0\,(s - s_0), \tag{10.8}$$

where $s$ and $s_0$ are the molar entropy values of the gas at pressure $p$ and $p_0$, respectively (vid. Eq. 7.3). We see in Eqs. 10.7 and 10.8 that exergy is zero $E = 0$ if $p = p_0$, while exergy is greater than zero $E > 0$ if $p \neq p_0$; i.e. expansion will occur, $p \rightarrow p_0$ or $p_0 \rightarrow p$. The pressure-dependent exergy of a gas have thus a positive value when the gas pressure is higher or lower than the atmospheric pressure: namely, a gas at a pressure higher or lower than the atmospheric pressure possesses an amount of exergy.

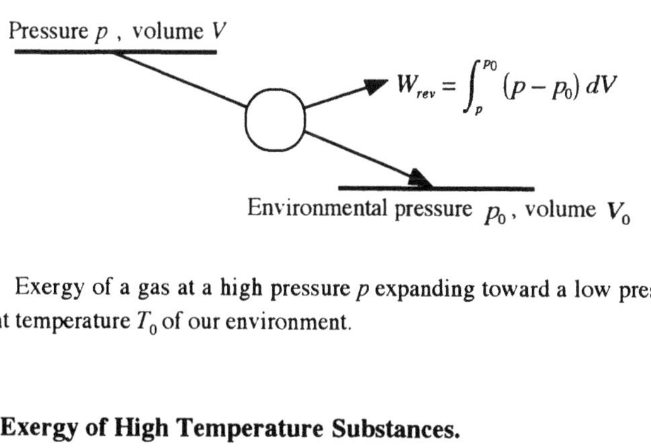


Fig. 10.2. Exergy of a gas at a high pressure $p$ expanding toward a low pressure $p_0$. at constant temperature $T_0$ of our environment.

## 10. 4. Thermal Exergy of High Temperature Substances.

Let us consider a certain substance at a high temperature $T$ from which a reversible heat engine receives an amount of heat $dQ(= dH)$ and performs an amount of reversible work $dW_{rev}$ releasing an amount of heat $dQ_0(= dH_0)$ into the environment at room temperature $T_0$ as shown in Fig. 10.3. The substance continues supplying heat to the engine until its temperature is equal to the environmental temperature $T_0$. During the heat supply $dQ$ the substance decreases its temperature by $dT$ as shown in Eq. 10.9 and decreases its exergy by $dE$ equal to the reversible work done, $dW_{rev}$, by the engine as shown in Eq. 10.10:

$$dQ = dH = C_p\, dT, \tag{10.9}$$

$$dE = dW_{rev} = \frac{T - T_0}{T}\, dQ > 0 \tag{10.10}$$

where $C_p$ is the heat capacity of the substance at constant pressure.

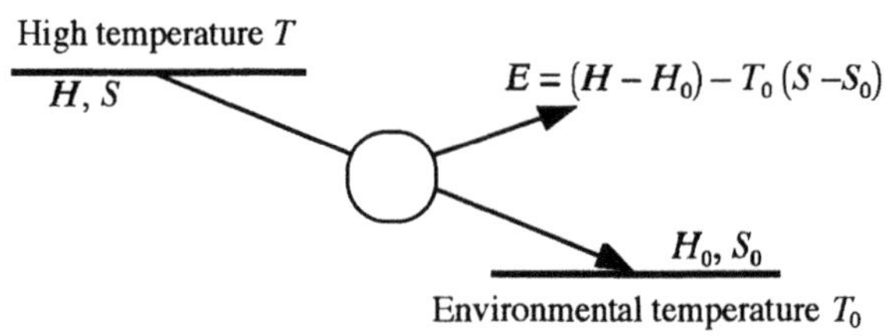


Fig. 10.3. Exergy of a substance at a high temperature $T$ releasing heat in decreasing its temperature from $T$ toward environmental temperature $T_0$ at atmospheric pressure.

The thermal exergy $E$ of a high temperature substance, which decreases its temperature from $T$ to $T_0$ at constant atmospheric pressure $p_0$, is hence given by Eq. 10.11:

$$E = W_{rev} = \int_{T_0}^{T} \frac{T - T_0}{T}\, dQ = \int_{T_0}^{T} \frac{T - T_0}{T}\, dH = \int_{T_0}^{T} C_p\left(1 - \frac{T_0}{T}\right) dT = \int_{T_0}^{T} C_p\, dT - T_0 \int_{T_0}^{T} \frac{C_p}{T}\, dT$$

$$= \int_{T_0}^{T} C_p\, dT - T_0 \int_{T_0}^{T} \frac{dQ}{T} = \int_{T_0}^{T} C_p\, dT - T_0 \int_{S_0}^{S} dS = (H - H_0) - T_0\,(S - S_0) > 0, \tag{10.11}$$

where $H$ and $H_0$ are the enthalpy of the substance at temperature $T$ and $T_0$; $S$ and $S_0$ are the entropy of the substance at temperature $T$ and $T_0$, respectively. We notice in Eqs. 10.10 and 10.11 that the exergy $E$ of a high temperature substances has always a positive quantity so that work can be obtained from it with respect to the environment.

Equation 10.11 yields the exergy $\varepsilon$ for one mole of the substance in terms of its molar enthalpy $h$ and molar entropy $s$ as shown in Eq. 10.12:

$$\varepsilon = (h - h_0) - T_0\,(s - s_0). \tag{10.12}$$

In fact, this equation 10.12 defines the thermal exergy of a high temperature substance.

When any phase transformation such as condensation or solidification occurs in the temperature fall, an additional exergy $\Delta\varepsilon$ of the latent heat of the phase transformation takes part in the available energy as shown in Eq. 10.13:

$$\Delta\varepsilon = q_t\left(1 - \frac{T_0}{T_t}\right), \tag{10.13}$$

where $q_t$ is the molar latent heat of the phase transformation at the transformation temperature $T_t$. On decreasing the temperature (condensation and solidification) the latent heat is negative $q_t < 0$; and hence if $T_t > T_0$, then an amount of exergy is released $\Delta\varepsilon < 0$. On increasing the temperature (melting and vaporization) the latent heat is positive $q_t > 0$; and hence if $T_t > T_0$, then an amount of exergy is taken up $\Delta\varepsilon > 0$; however, if $T_t < T_0$, then an amount of exergy is released $\Delta\varepsilon < 0$. The numerical values of the molar exergy of substances and of their phase transformations are tabulated in the literature on engineering thermodynamics [Refs. 9 and 11.].

The fall in temperature from $T$ to the environmental temperature $T_0$ releases an amount of enthalpy $(h - h_0)$, of which however only the exergy part can be used for the available and transformable energy to obtain useful work or products by means of reversible processes. We then define the *energy availability* $\lambda$ of a high temperature substance as in Eq. 10.14:

$$\lambda = \frac{\varepsilon}{h - h_0}. \tag{10.14}$$

The molar exergy of a high temperature substance in Eq. 10.11, if no phase transformation is involved in the temperature range considered, can be expressed by Eq. 10.15 in terms of the mean molar heat capacity $c_p^{\text{mean}}$:

$$\varepsilon = \int_{T_0}^{T} c_{p,h}^{\text{mean}}\, dT - T_0\, c_{p,s}^{\text{mean}} \int_{T_0}^{T} \frac{1}{T}\, dT = c_{p,h}^{\text{mean}}(T - T_0) - c_{p,s}^{\text{mean}}\, T_0 \ln\frac{T}{T_0} \tag{10.15}$$

$$c_{p,h}^{mean} = \frac{\int_{T_0}^{T} c_p\, dT}{T - T_0}, \qquad c_{p,s}^{mean} = \frac{\int_{T_0}^{T} \frac{c_p}{T}\, dT}{\ln T - \ln T_0}$$

where $c_p$ is the molar heat capacity of the substance and is usually a function of temperature. We normally assume that $c_{p,h}^{mean}$ is equal to $c_{p,s}^{mean}$; $c_{p,s}^{mean} = c_{p,h}^{mean} = c_p^{mean}$. This molar exergy is compared with the molar enthalpy given by Eq. 10.16:

$$h - h_0 = \int_{T_0}^{T} c_p^{\text{mean}}\, dT = c_p^{\text{mean}}(T - T_0). \tag{10.16}$$

We thus obtain the energy availability $\lambda$ of the thermal energy of a substance at a high temperature $T$ as shown in Eq. 10.17:

$$\lambda = \frac{\varepsilon}{h - h_0} = 1 - \frac{T_0}{T - T_0}\ln\frac{T}{T_0}. \tag{10.17}$$

This equation however is valid only when no phase transformation is involved.

**10. 5. Thermal Exergy of Low Temperature Substances.**

If the temperature $T$ of a substance is lower than the temperature $T_0$ of the environment, a heat engine can be operated between the environment (heat source) and the low temperature substance (heat sink) . Let us consider a reversible heat engine as shown in Fig 10.4 in which the engine gas receives an amount of heat $dQ$ from the environment at atmospheric temperature $T_0$ and performs an amount of reversible work $dW_{rev}$ releasing an amount of heat into the low temperature substance at temperature $T$, whose enthalpy is then increased by an amount $dH = dQ - dW_{rev} > 0$. From the efficiency of the reversible engine we have Eq. 10.18:

$$dW_{rev} = \frac{T_0 - T}{T_0} dQ = \frac{T_0 - T}{T_0} (dH + dW_{rev}), \qquad dW_{rev} = \frac{T_0 - T}{T} dH > 0. \qquad (10.18)$$

The reversible heat engine continues operating until the temperature of the substance becomes equal to the atmospheric temperature. The thermal exergy $E$ of the substance at a low temperature $T$ is thus given by Eq. 10.19:

$$E = W_{rev} = \int_T^{T_0} \frac{T_0 - T}{T} dH = -\int_T^{T_0} dH + T_0 \int_T^{T_0} \frac{dH}{T} = H - H_0 - T_0 (S - S_0) > 0, \quad (10.19)$$

This equation indicates that the thermal exergy of a low temperature substance has a positive sign and hence that the low temperature substance possesses an amount of available exergy relative to the same substance at the atmospheric reference temperature.

The molar exergy of a low temperature substance is then given by Eq. 10.20:

$$\varepsilon = h - h_0 - T_0 (s - s_0), \qquad (10.20)$$

which is in the same form as Eq. 10.13 for a high temperature substance. With low temperature substances $(h - h_0) < 0$ and $(s - s_0) < 0$, but the balance, as shown in Eq. 10.18, gives us $\varepsilon > 0$ as a whole indicating that they can release exergy to do an amount of work.

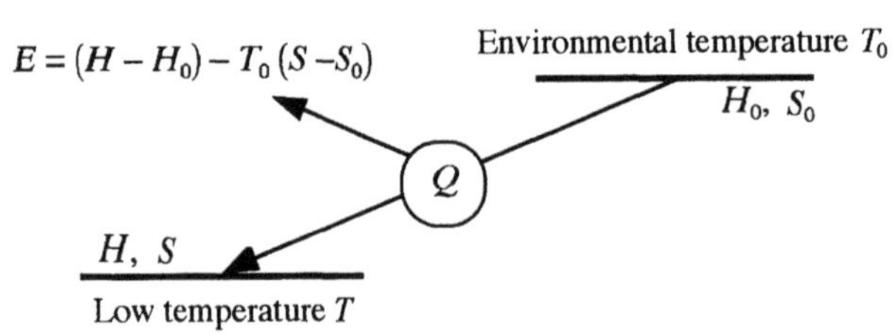


Fig. 10.4. Exergy of a substance at a low temperature receiving heat and increasing its temperature toward environmental temperature $T_0$.

**10. 6. Exergy in Mixing Substances.**

In the foregoing section 5.10 we have discussed the thermodynamic functions of mixing which arise in the mixing of substances. At temperature $T_0$ and pressure $p_0$ the exergy $\varepsilon^M$ due to the mixing of substances for one mole of the mixture is given by the free enthalpy $g^M = h^M - T_0\, s^M$ for the mixing of substances, where $h^M$ and $s^M$ are the enthalpy and the entropy for the mixing of substances for one mole of the mixture; $h^M$ is zero if the mixture is a perfect system defined by the symmetrical reference system shown in section 8.3, such as an ideal gas mixture and a perfect solution. For a perfect mixture consisting of molar fraction $x_i$ of each of the substances present, the entropy of mixing $s^M$ for one mole of the mixture is given, from Eq. 5.45, by Eq. 10.21:

$$s^M = -\sum_i x_i\left(s_i^{mixture} - s_i^{pure}\right) = -\sum_i x_i\, R \ln x_i\,. \qquad (10.21)$$

where $s_i^{pure}$ and $s_i^{mixture}$ are the partial molar entropy of each of the substances $i$ in the pure state and in the mixture, respectively. The molar exergy for the mixing of substances in the ideal mixture at temperature $T_0$ is then expressed by Eq. 10.22 :

$$\varepsilon^M = h^M - T_0\, s^M = h^M + \sum_i x_i\, T_0\, R \ln x_i \approx \sum_i x_i\, T_0\, R \ln x_i\,. \qquad (10.22)$$

As mentioned in sections 5.9 and 8.4 the enthalpy of mixing is zero $h^M = 0$ for perfect mixtures, while it is not zero $h^M \neq 0$ for non-ideal mixtures. In the case of an ordinary gas mixture we may assume $h^M = 0$.

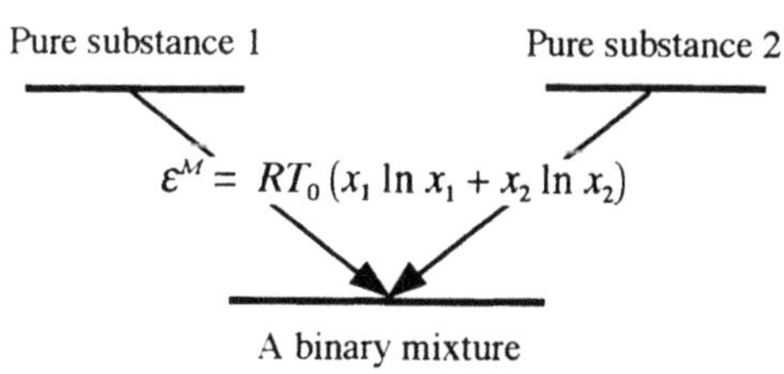


Fig. 10.5. Exergy of mixing for one mole of a binary ideal gas mixture at our environmental temperature $T_0$ and pressure $p_0$.

The molar exergy $\varepsilon$ of an ideal gas mixture ($h^M = 0$) at the environmental temperature $T_0$ and pressure $p_0$ is hence obtained from Eq. 10.22 as given by Eq. 10 23:

$$\varepsilon = \sum_i x_i\, \varepsilon_i^{pure} + \varepsilon^M = \sum_i x_i\, \varepsilon_i^{pure} + h^M - T_0\, s^M$$

$$= \sum_i x_i\, \varepsilon_i^{pure} + \sum_i R\, T_0\, x_i \ln x_i, \qquad (10.23)$$

where $\varepsilon_i^{pure}$ is the molar exergy of each of the pure substances *i*. In Eq. 10 23 the first term is the unitary quantities of exergy and the second term is the quantities of exergy of mixing. We notice that the exergy of mixing $\varepsilon^M$ has always a negative sign and hence that mixing reduces exergy of the system. Equation 10.23 can apply not only to gaseous ideal mixtures but also to perfect solutions in condensed phases (liquid or solid). For dilute-ideal and non-ideal solutions however the enthalpy of mixing is not zero $h^M \neq 0$, and hence it has to be taken into account in evaluating the exergy of mixing. Figure 10.5 shows the molar exergy of mixing in a binary ideal gas mixture at the environmental temperature $T_0$ and pressure $p_0$.

Consider now at the environmental conditions of temperature $T_0$ and pressure $p_0$ a gas mixture whose compositional values $x_i$ are different from those values $x_{i,0}$ of the stable atmospheric air. For this stable atmospheric air each of the partial molar exergy values of the constituents are zero. From Eq. 10.23 we then obtain the exergy of the gas mixture as shown in Eq. 10.24:

$$\varepsilon = \sum_i R\,T_0\,x_i \ln \frac{x_i}{x_{i,0}}\,. \tag{10.24}$$

A gas of oxygen-rich air of the composition $x_{O_2} = 0.30$ and $x_{N_2} = 0.70$, for example, possesses the exergy for one mole shown in Eq. 10.25 relative to the atmospheric air of the composition $x_{O_2,0} = 0.21$ and $x_{N_2,0} = 0.79$:

$$\varepsilon = R\,T_0\left(0.30 \ln \frac{0.30}{0.21} + 0.70\ \ln \frac{0.70}{0.79}\right) = 0.0223\,R\,T_0\,. \tag{10.25}$$

Furthermore, pure oxygen gas, whose molar entropy is lower by an amount of $R\ln 0.2034$ than that of gaseous oxygen molecules in the atmospheric wet air ($x_{O_2} = 0.2034$) at the standard temperature and pressure, possesses its *standard molar exergy* $\varepsilon^0_{O_2}$ given by Eq. 10.26 relative to the atmospheric wet air:

$$\varepsilon^0_{O_2} = RT_0 \ln\left(\frac{1}{x_{i,0}}\right) = -\,298.2 \times R \times \ln 0.2034 = 3.97\ \mathrm{J{\cdot}mol^{-1}}, \tag{10.26}$$

where the superscript 0 indicates a pure substance at the standard state (pressure 101.3 kPa, temperature 298.2 K) and subscript 0 indicates the exergy reference state.

In the case of non-ideal mixtures (e.g. liquid and solid solutions), the activity $a_i = \gamma_i x_i$ has to be used instead of the molar fraction $x_i$ of substance *i* after the logarithmic sign in Eq. 10.23 to express the mixing term of the exergy at the exergy reference temperature $T_0$ and pressure $p_0$ as shown in Eq. 10.27:

$$\varepsilon(a_i) = \sum_i x_i\,\varepsilon_i^*(a_i{=}1) + \sum_i R\,T_0\,x_i \ln a_i = \sum_i x_i\,\varepsilon_i^*(a_i{=}1) + \sum_i R\,T_0\,x_i \ln \gamma_i\,x_i, \tag{10.27}$$

where $\varepsilon_i^*(a_i{=}1)$ is the unitary exergy at unit activity of *i* and $\gamma_i$ is the activity coefficient of *i*. The unitary exergy $\varepsilon_i^*$ is equivalent to the exergy of each of the pure substances *i* in the mixture when the reference system for the unitary quantities is symmetrical (vid. section 8.3).

When the mixture is at a temperature $T$ different from the reference temperature $T_0$, the exergy $\varepsilon$ for one mole of a non-ideal mixture is given by Eq. 10.28:

$$\varepsilon(T, a_i) = \sum_i x_i \, \varepsilon_i^*(T, a_i{=}1) + \sum_i R\, T_0\, x_i \ln a_i - \sum_i R\, T(T - T_0)\, x_i \left(\frac{\partial \ln a_i}{\partial T}\right)_{p,\, x_i}, \qquad (10.28)$$

where the third term is related with the temperature-dependence of the activity.

## 10. 7. Chemical Exergy of Substances.

A chemical substance has its chemical energy in terms of the chemical potential and has its chemical exergy as well. Let us consider a chemical substance present at unit activity in the normal environment at temperature $T_0$ and pressure $p_0$ and examine its chemical exergy in relation with the exergy reference species in the atmospheric air, in seawater, and in lithospheric solids (Refs. 9 and 11).

Typical exergy reference species in the atmospheric air are oxygen $O_2$ ($x_{o_2}$ = 0.21 in air), nitrogen $N_2$ ($x_{N_2}$ = 0.78 in air), carbon dioxide $CO_2$, and water vapor $H_2O$ at their respective concentrations. The chemical exergy $\varepsilon^0_{O_2}$ of oxygen $O_2$ at the unit activity (unit fugacity or unit atmospheric pressure), then, arises from its mixing exergy as shown by Eq. 10.26 in the preceding section: we call it the *standard molar chemical exergy* of pure oxygen gas:

$$\varepsilon^0_{O_2} = -R\, T_0 \ln x_{O_2,0}, \qquad (10.29)$$

where $x_{O_2,0}$ is the molar fraction of oxygen moles in the atmospheric air.

We next consider metallic iron whose exergy reference species are oxygen molecules in the atmospheric air and solid iron oxide $Fe_2O_3$, which is the most stable existence of iron in the top layer of the lithosphere. In the atmospheric air metallic iron reacts with oxygen gas to form iron oxide (corrosion of metallic iron). The reaction at the standard state (unit activity, standard pressure 101.3 kJ, and standard temperature 298 K) is expressed in Eq. 10.30:

$$2\,Fe_{(solid)} + \frac{3}{2}\,O_{2(gas)} \rightarrow Fe_2O_{3(solid)}, \qquad A^0_{Fe_2O_3} = 742.6\ \mathrm{kJ\cdot mol^{-1}}, \qquad (10.30)$$

where $A^0_{Fe_2O_3}$ is the standard affinity of the reaction and, as described in chapter 4 and 5, it is expressed by the difference in the standard free enthalpy, $G^0$, between the product and the reactants; $A^0_{Fe_2O_3} = -\Delta G^0 = -(\Delta H^0 - T_0\Delta S^0) = 2\mu^0_{Fe} + (3/2)\mu^0_{O_2} - \mu^0_{Fe_2O_3}$ with $\mu^0_i$ being the standard chemical potential of species $i$. If advancing reversibly in the standard state, the reaction holds the following exergy balance:

$$2\varepsilon^0_{Fe} + \frac{3}{2}\varepsilon^0_{O_2} = \varepsilon^0_{Fe_2O_3} + A^0_{Fe_2O_3}, \qquad (10.31)$$

where $\varepsilon^0_{Fe_2O_3}$ is zero with $Fe_2O_3$ being one of the exergy reference species and $\varepsilon^0_{O_2}$ is given by Eq. 10.26. We, hence, obtain the standard exergy of metallic iron as shown in Eq. 10.32:

$$\varepsilon^0_{Fe} = \frac{1}{2}\left\{0 + 742.6 - \left(\frac{3}{2}\right) \times 3.97\right\} = 368.3 \text{ kJ·mol}^{-1} \tag{10.32}$$

In the same way as Eqs. 10.31 and 10.32, we also obtain the standard exergy of gaseous hydrogen and that of solid carbon, $\varepsilon^0_{H_2}$ and $\varepsilon^0_{C}$, as shown in Eqs. 10.33 and 10.34, respectively:

$$\varepsilon^0_{H_2} + \frac{1}{2}\varepsilon^0_{O_2} = \varepsilon^0_{H_2O} + A^0_{H_2O}, \tag{10.33}$$

$$\varepsilon^0_{C} + \varepsilon^0_{O_2} = \varepsilon^0_{CO_2} + A^0_{CO_2}, \tag{10.34}$$

where $A^0_{H_2O}$ is the standard affinity of the reaction, $H_{2,(gas)} + 0.5O_{2,(gas)} = H_2O_{(liquid)}$; $A^0_{CO_2}$ is the standard affinity of the reaction, $C_{(solid)} + O_{2,(gas)} = CO_{2,(gas)}$; $\varepsilon^0_{H_2O}$ is the standard molar exergy of liquid water $H_2O$, $\varepsilon^0_{H_2O} = -RT_0 \ln x_{H_2O,0}$; and $\varepsilon^0_{CO_2}$ is the standard molar exergy of $CO_2$, $\varepsilon^0_{CO_2} = -RT_0 \ln x_{CO_2,0}$, with $x_{H_2O,0}$ and $x_{CO_2,0}$ being the molar fractions of $H_2O$ and $CO_2$ in the wet atmospheric air, respectively.

From the foregoing discussion, it follows that the standard exergy of one of the reactants can be estimated by use of the standard affinity of the reaction, provided that we know the values of the standard exergy of the other reactants and products. The numerical values of the molar exergy thus obtained of various chemical substances in the standard state (temperature $T^0 = 298$ K, pressure $p^0 = 101.3$ kPa, activity $a^0_i = 1$) are tabulated as the standard chemical exergy $\varepsilon^0_i$ of chemical substances in the literature on engineering thermodynamics [Ref. 9 .].

Furthermore, in analogy to the partial molar quantities of thermodynamic functions, the *partial molar chemical exergy*, $\varepsilon_{chem,i}$, can be defined for a substance $i$ in a gaseous mixture, in a liquid solution, and in a solid solution as shown in Eq. 10.35:

$$\varepsilon_{chem,i} = \left(\frac{\partial E_{chem}}{\partial n_i}\right)_{T_0, p_0, n_{j(j\neq i)}}, \tag{10.35}$$

where $E_{chem}$ is the total chemical exergy of the mixture or solution under consideration at the environmental temperature $T_0$ and pressure $p_0$. The partial molar chemical exergy of substance $i$ at temperature $T_0$ can be expressed in terms of the standard molar chemical exergy $\varepsilon^0_{chem,i}$ and the activity $a_i$ of the substance $i$ as follows:

$$\varepsilon_{chem,i} = \varepsilon^0_{chem,i} + RT_0 \ln a_i \,. \tag{10.36}$$

Then, the exergy $E_{chem}$ of the mixture at the standard temperature and pressure is obtained:

$$E_{chem} = \sum_i n_i\, \varepsilon_{chem,i} = \sum_i n_i\, \varepsilon^0_{chem,i} + RT_0 \sum_i n_i \ln a_i. \tag{10.37}$$

This equation 10.37 is equivalent to Eq. 10.27.

We now consider a chemical reaction that occurs at a temperature $T$ and pressure $p$ other than the standard state ($T^0$, $p^0$, and unit activity):

$$R \rightarrow P, \qquad \text{reaction affinity } A \tag{10.38}$$

where R denotes the reactants, P denotes the products, and $A$ is the affinity of the chemical reaction at temperature $T$. The reaction exergy, $\Delta E_{chem,T,p}$, dissipated during the reaction is equivalent to the reaction affinity and is given by Eq. 10.39 and in Fig. 10.6:

$$\Delta E_{chem,T,p} = E_{react,T,p} - E_{prod,T,p} = A, \tag{10.39}$$

where $E_{react,T,p}$ and $E_{prod,T,p}$ are the amounts of exergy of the reactants and products, respectively. The reaction affinity can be estimated thermodynamically so that by calculation we obtain the change in exergy associated with the chemical reaction.

The reaction exergy is usually assumed to consist of a chemical part at the standard state ($T^0$, $p^0$, and unit activity) and a physical part associated with the physical state of the reaction. The chemical part $\Delta E^0_{chem,T^0,p^0}$ is equivalent to the standard affinity $A^0$ of the reaction, and the physical part $\Delta E_{phy}$ is due to the change in temperature, pressure, and concentration of the reactants and products between the standard state and the state at which the reaction proceeds:

$$\Delta E_{chem,T,p} = \Delta E^0_{chem,T^0,p^0} + \Delta E_{phy} = A^0 + \Delta E_{phy}. \tag{10.40}$$

This is a general expression for the reaction exergy of a chemical reaction taking place at conditions other than the standard state.

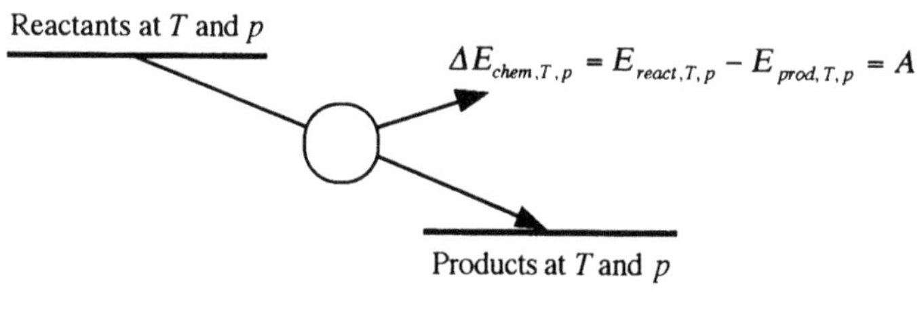


Fig. 10.6. Chemical exergy $\Delta E_{chem,T,p}$ associated with a chemical reaction at temperature $T$ and pressure $p$: $E_{react,T,p}$ and $E_{prod,T,p}$ are the amounts of exergy of the reactants and products; $A$ is the affinity of the reaction.

## 10. 8. Standard Chemical Exergy of Substances.

In chemical thermodynamics the standard chemical potential $\mu_i^0$ of a compound $i$ is defined as the molar free enthalpy $\Delta g_f^0$ for the formation of the compound from its constituent elements $j$ in their stable molecular form in the standard state, and their chemical potential values are set zero in the standard state: $\mu_i^0 = \Delta g_f^0$. In exergy engineering the standard molar exergy $\varepsilon_i^0$ of a compound $i$ is defined as consisting of the molar free enthalpy $\Delta g_f^0$ for the formation of the compound in the standard state from its constituent elements and the stoichiometrical sum of the standard chemical exergy values $\varepsilon_j^0$ of the constituent elements $j$ in their stable state at the standard temperature $T^0$ and pressure $p^0$: $\varepsilon_i^0 = \Delta g_f^0 + \sum_j \nu_j \varepsilon_j^0$.

In calculating the numerical values of the standard molar exergy $\varepsilon_i^0$ of chemical elements and compounds, we usually make clear the exergy reference species at zero level of exergy in our natural environment of the atmosphere, the hydrosphere and the lithosphere.

Table 10.1. Standard molar chemical exergy of a few substances relative to the reference species in the atmosphere [Refs. 9 and 11.].

| Substance | $O_2$ Gas | $N_2$ Gas | $CO_2$ Gas | $H_2O$ Gas | $H_2$ Gas | C Solid |
|---|---|---|---|---|---|---|
| $\varepsilon_i^0/\mathrm{kJ \cdot mol^{-1}}$ | 3.97 | 0.72 | 19.89 | 9.5 | 236.09 | 410.26 |
| Reference substance in the atmosphere | $O_2$ | $N_2$ | $CO_2$ | $H_2O_{(gas)}$ | $H_2O_{(gas)}$ $O_2$ | $CO_2$ $O_2$ |

Atmospheric $H_2O$ is in equilibrium with the ocean.

For gaseous substances such as oxygen $O_2$, nitrogen $N_2$, carbon dioxide $CO_2$, and water vapor $H_2O$, the standard chemical exergy can be calculated from their concentrations in the atmospheric air by use of Eq. 10.26. Furthermore, the standard affinity $A^0$ of the reaction, $H_{2(gas)} + (1/2)O_{2(gas)} = H_2O_{(gas)}$, yields through Eq. 10.33 the standard chemical exergy of gaseous hydrogen $H_2$. Similarly, the standard affinity $A^0$ of the reaction, $C_{(solid)} + O_{2(gas)} = CO_{2(gas)}$, yields the standard chemical exergy of solid carbon C. Table 10.1 shows the standard molar exergy of these substances relative to the exergy reference species in the atmospheric air at temperature 298 K and pressure 101.3 kPa.

For substances which are not present in the atmosphere but in the ocean, we can take the reference species of zero exergy level at the most stable state of their existence in seawater. For example, metallic sodium takes its reference level at the state of sodium ions in seawater and the standard chemical exergy $\varepsilon_{Na}^0$ of metallic sodium is equivalent to the free enthalpy required to form solid sodium from sodium ions in seawater: $Na^{+}_{sea} + 0.5H_{2(gas)} = Na_{solid} + H^{+}_{sea}$. Taking account of Eq. 9.38 for the chemical potential of a hydrated ion, we obtain the standard exergy of solid sodium $\varepsilon^{0}_{Na}$ from the free enthalpy change for the formation of solid sodium from sodium ions in seawater as shown in Eq. 10.41:

$$\varepsilon^{0}_{Na} = -\mu^{0}_{Na^{+}} - R\,T_0 \ln c_{Na^{+}}\,\gamma_{Na^{+}} + \frac{1}{2}\,\varepsilon^{0}_{H_2} - 2.303\,R\,T_0\,(pH)_{sea}, \tag{10.41}$$

where $\mu^{0}_{Na^{+}}$ is the standard chemical potential of hydrated sodium ions, $\varepsilon^{0}_{H_2}$ is the standard molar exergy of gaseous hydrogen, $(pH)_{sea}$ is the pH of seawater, and where $c_{Na^{+}}$ and $\gamma_{Na^{+}}$ are the concentration and the activity coefficient of sodium ions in seawater, respectively. Table 10.2 shows the numerical values of the standard chemical exergy of a few substances relative to their ionic form present in the ocean at the standard temperature and pressure.

Table 10.2. Standard molar chemical exergy of a few substances relative to the ocean. (Ref. 11)

| Substance | Na<br>Solid | P<br>Solid | $Cl_2$<br>Gas | S<br>Solid | Zn<br>Solid | As<br>Solid |
|---|---|---|---|---|---|---|
| $\varepsilon_i^0/kJ\cdot mol^{-1}$ | 335.9 | 865.3<br>866.0 | 124.1 | 607.3 | 338.7 | 486.1<br>476.5 |
| Reference substance in the ocean | $Na^{+}$ | $H_2PO_4^{-}$<br>$HPO_4^{-}$ | $Cl^{-}$ | $SO_4^{2-}$ | $Zn^{2+}$ | $HAsO_4^{2-}$<br>$HAsO_4^{-}$ |

Table 10.3. Standard molar chemical exergy of a few substances relative to the lithosphere (Ref. 11 )

| Substance | Fe | $Fe_2O_3$ | $Fe_3O_4$ | Si | $SiO_2$ | MnO |
|---|---|---|---|---|---|---|
| $\varepsilon_i^0/kJ\cdot mol^{-1}$ | 368.41 | 0 | 96.97 | 800.3 | 0 | 100.36 |
| Reference substance on the earth | $Fe_2O_3$ | $Fe_2O_3$ | $Fe_2O_3$ | $SiO_2$ | $SiO_2$ | $MnO_2$ |

In the case of solid substances the reference species is often set at the most stable solid compounds in lithospheric rocks. For example, metallic iron is most stable in the form of its oxides. The standard chemical exergy of metallic iron can then be obtained from the standard affinity $A_f^0$ of the formation of iron oxide, $Fe + 0.75O_2 = 0.5Fe_2O_3$: $A_f^0 = \varepsilon_{Fe}^0 + 0.75\varepsilon_{O_2}^0 - 0.5\varepsilon_{Fe_2O_3}^0$ and $\varepsilon_{Fe_2O_3}^0 = 0$; hence $\varepsilon_{Fe}^0 = A_f^0 - 0.75\varepsilon_{O_2}^0$. Table 10.3 shows the standard molar chemical exergy of a few substances relative to the solid reference species in the lithosphere at the standard temperature and pressure.

## 10. 9. Total Exergy of Substances.

The exergy of substances is arranged as consisting of a physical part and a chemical part. The chemical part of exergy is conventionally attributed to the chemical formation of the substances in the standard state from the exergy reference level substances in our environment, while the physical part of exergy is attributed to the changes in temperature, pressure and concentration (mixing) of the substances. The overall exergy $E$ of a gaseous mixture of substances is thus expressed as shown in Eq. 10.42:

$$E = \sum_i n_i\varepsilon_i^0 + RT_0\sum_i n_i \ln\frac{p_i}{p_0} + \sum n_i c_{p,i}^{\text{mean}}\left(T - T_0 - T_0\ln\frac{T}{T_0}\right) + RT_0\sum_i\left\{n_i \ln\left(\frac{n_i}{\sum_i n_i}\right)\right\}. \qquad (10.42)$$

In this equation the first term on the right hand side is the chemical exergy (vid. Eqs. 10.29 and 10.32), the second term is the pressure exergy for gaseous substances (vid. Eq. 10.7), the third term is the thermal exergy due to the change in temperature (vid. Eqs. 10.11 and 10.15), the forth term is the mixing exergy due to the change in concentration of the substances (vid. Eq. 10.21). For mixtures in condensed phases (liquid or solid), the pressure exergy may be approximated by $V_m(p - p_0)$, where $V_m$ is the volume of the condensed phase at temperature $T$.

From the foregoing discussion in this chapter we realize that all the terms in Eq. 10.42 can be expressed in the form $(H - H_0) - T_0(S - S_0)$. We can therefore express the total exergy in a simple form as follows:

$$E = (H - H_0) - T_0(S - S_0) \qquad (10.43)$$

This is a *general expression* for the total exergy of substances.

## 10. 10. Exergy and Affinity

Exergy of a substance represents the level of available energy of the substance relative to the exergy reference zero level of the substance in our normal environment: the substance undergoes a spontaneous change in the direction in which its exergy decreases as the change proceeds. On the other hand, the affinity of a process, as mentioned in chapters 3 and 4, represents the difference in the available energy level (free enthalpy, Gibbs energy) between the initial state of the substance (reactant) and the final state of the substance (product); the process spontaneously proceeds in the direction in which its affinity decreases. This situation is illustrated in Fig. 10.7. We then realize that exergy and affinity resemble each other in that the two quantities represent available energy and determine the possibility and capacity of irreversible processes.

If we consider a physicochemical process in which reactant substances change from their initial state to their final state of product substances which is in equilibrium with the exergy-reference substances in the normal environment, the affinity of the process is represented in general by Eq. 10.43 in terms of enthalpy and entropy:

$$A = \left(H^{I}_{T,p} - H^{F}_{T_0,p_0}\right) - T_0\left(S^{I}_{T,p} - S^{F}_{T_0,p_0}\right), \tag{10.43}$$

where $H^{I}_{T,p}$ and $S^{I}_{T,p}$ are the enthalpy and the entropy of the initial state of the substances under consideration at temperature $T$ and pressure $p$; $H^{F}_{T_0,p_0}$ and $S^{F}_{T_0,p_0}$ are the enthalpy and the entropy of the final state of the substances in equilibrium with the exergy-reference substances in the normal environment at temperature $T_0$ and pressure $p_0$, respectively. This equation 10.43 is, in fact, equivalent to the general equation of exergy of the initial state of the substances under consideration as shown in Fig. 10.8:

$$E = \left(H^{I}_{T,p} - H^{F}_{T_0,p_0}\right) - T_0\left(S^{I}_{T,p} - S^{F}_{T_0,p_0}\right) = A. \tag{10.44}$$

We then realize that exergy is in principle the same energy function as affinity.

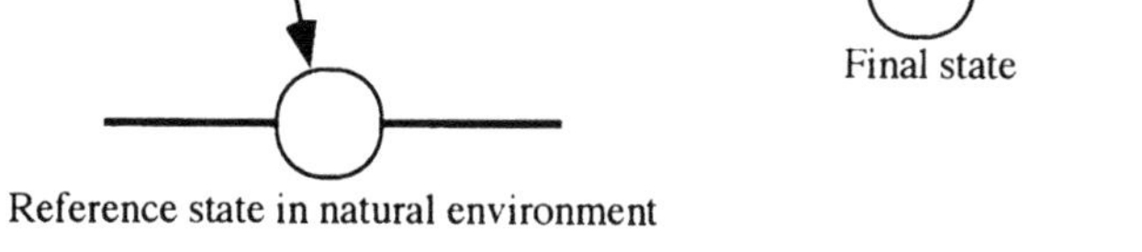


Fig. 10.7. Exergy and affinity

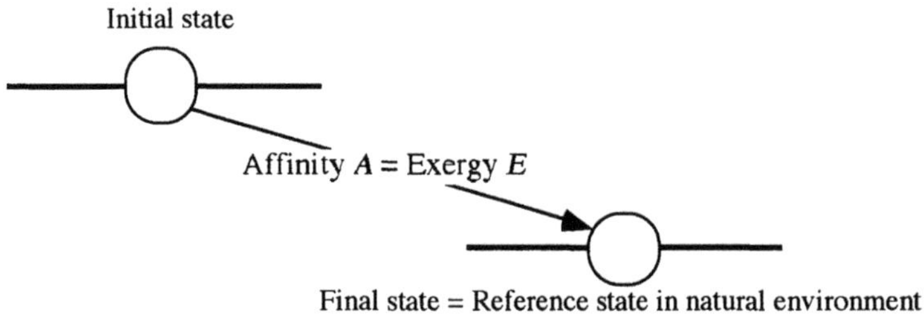


Fig. 10.8. Exergy equivalent to affinity when the final state of reaction is in equilibrium with the exergy-reference state.

# CHAPTER 11

# EXERGY DIAGRAM

The effective use of energy in a chemical plant results from employing the combination of processes that consumes the least amount of exergy as a whole in manufacturing the products. To examine the flow of exergy the diagram that visualizes the exergy balance in manufacturing processes has practically been used for the purpose of improving the exergy efficiency in the plants. Furthermore, a novel concept of exergy vectors has recently been brought forth to evaluate the exergy efficiency of various processes on an enthalpy-exergy diagram. By means of exergy vectors the minimum exergy thermodynamically required for a manufacturing process may be estimated in the diagram. This chapter describes the principle of exergy vector diagrams and their application to basic processes such as heating-cooling, compression-expansion, separation-mixing, and chemical reactions.

### 11. 1. Efficiency in the Use of Energy.

In chemical manufacturing processes there are a flow of substances (materials) $\Sigma n_i$, a flow of work $W$, and a flow of heat $Q$, which are all accompanied by a flow of exergy $E$. The inlet flow of exergy $dE_+/dt$ (positive sign) into and the outlet flow of exergy $dE_-/dt$ (negative sign) out of an open system of chemical processes at a temperature $T$ are given by Eqs. 11.1 and 11.2, respectively:

$$\frac{dE_+}{dt} = \sum_{i+} \frac{dn_{i+}}{dt}\,\varepsilon_{i+} + \frac{dW_+}{dt} + \left(\frac{T - T_0}{T}\right)\frac{dQ_+}{dt}, \qquad (11.1)$$

$$\frac{dE_-}{dt} = \sum_{i-} \frac{dn_{i-}}{dt}\,\varepsilon_{i-} + \frac{dW_-}{dt} + \left(\frac{T - T_0}{T}\right)\frac{dQ_-}{dt}, \qquad (11.2)$$

where $n_i$ is the number of moles, $\varepsilon_i$ is the molar exergy of substance $i$, and $T_0$ is the exergy reference temperature. On the right hand side in Eqs. 11.1 and 11.2 the first term is the exergy flow of substances, the second term is the exergy flow of work, and the third term is the exergy flow of heat. Figure 11.1 shows the flow of exergy through an open system at a temperature $T$ in which physical and chemical processes are taking place in irreversible modes. Since exergy decreases with the advancement of irreversible processes in the system, the outlet flow $dE_-/dt$ is less than the inlet flow $dE_+/dt$ of exergy:

$$\frac{dE_+}{dt} \geq \frac{dE_-}{dt}, \qquad \frac{dE_+}{dt} = \frac{dE_-}{dt} + T_0 \frac{dS_{irr}}{dt} = \frac{dE_-}{dt} + \frac{dE_{loss}}{dt}, \tag{11.3}$$

where $T_0$ is the exergy reference temperature of 298 K, $dS_{irr}/dt$ is the creation of entropy, and $T_0\,(dS_{irr}/dt) = dE_{loss}/dt$ is the loss of exergy due to the irreversible processes occurring in the system.

Equation 11.3 is occasionally called *the law of exergy loss* or the Gouy–Stodola's relation (G. Gouy and A. Stodola) who first discovered independently of each other in the late nineteenth century the loss of maximum work due to the irreversibility of thermal processes [Ref. 9.].

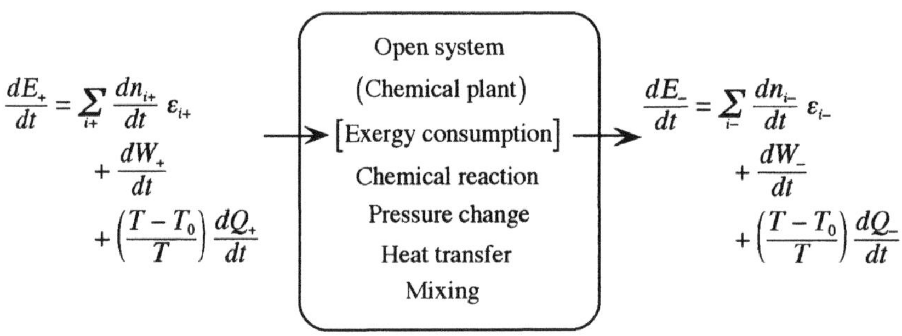


Fig. 11.1. Flow of exergy through an open system at temperature $T$.

The efficiency of exergy utilization in an open system (chemical reactor) is then defined by Eq. 11.4:

$$\eta = \frac{\dfrac{dE_-}{dt}}{\dfrac{dE_+}{dt}} = 1 - \frac{\dfrac{dE_{loss}}{dt}}{\dfrac{dE_+}{dt}}, \tag{11.4}$$

where $dE_{loss}/dt$ is the rate of the loss of exergy (consumption of exergy) due to the irreversible processes occurring in the open system: this is also called the rate of *internal exergy loss* in the system. Reducing the loss of exergy obviously leads to improving the exergy utilization in chemical plants.

**11. 2. Exergy Balance Diagram.**

Let us consider a physicochemical process occurring in an irreversible way in a chemical reactor, which receives raw materials with enthalpy $H_1$ and entropy $S_1$ together with an amount of heat $Q_1$ at temperature $T_1$; and generates a useful product expressed by enthalpy $H_{prod}$ and entropy $S_{prod}$ giving off enthalpy $H_2$ and entropy $S_2$ as effluent materials, and discharging an amount of waste heat $Q_0$ at the environmental temperature $T_0$ as shown in Fig. 11.2. The product may be useful work or useful substances, both being represented in terms of $H_{prod}$ and $S_{prod}$. The effluent materials may be waste or reusable. For the energy balance (the enthalpy balance) of the process we have Eq. 11.5:

$$H_1 + Q_1 = H_2 + Q_0 + H_{prod} . \qquad (11.5)$$

If the process occurs in a reversible way instead of an irreversible way, a greater amount of products, expressed by $H_{prod,rev}$ and $S_{prod,rev}$, and a smaller amount of waste heat, $Q_{0,rev}$, will be yielded than those in the irreversible process; the energy balance is then given by Eq. 11.6:

$$H_1 + Q_1 = H_2 + Q_{0,rev} + H_{prod,rev} . \qquad (11.6)$$

From Eqs. 11.5 and 11.6 we obtain Eq. 11.7:

$$H_{prod,rev} - H_{prod} = Q_0 - Q_{0,rev} , \qquad (11.7)$$

which indicates that the decreased amount of products in the irreversible process compared to the reversible process is accompanied with an increased amount of waste heat.

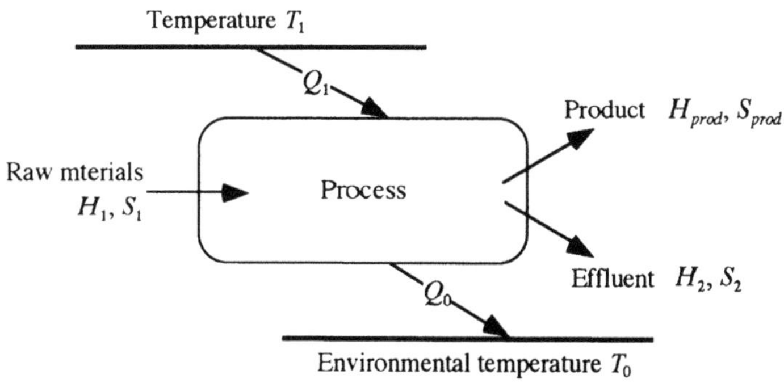


Fig. 11.2. Physicochemical irreversible flow process.

In the reversible process no entropy creation occurs:

$$\frac{Q_1}{T_1} + S_1 = \frac{Q_{0,rev}}{T_0} + S_2 + S_{prod,rev}, \tag{11.8}$$

whereas, in the irreversible process an amount of entropy creation $\Delta S_{irr}$ occurs as shown in Eq. 11.9:

$$\frac{Q_1}{T_1} + S_1 + \Delta S_{irr} = \frac{Q_0}{T_0} + S_2 + S_{prod}. \tag{11.9}$$

We then obtain Eqs. 11.10 and 11.11 from Eqs. 11.7, 11.8 and 11.9:

$$Q_0 - Q_{0,rev} = T_0(\Delta S_{irr} + S_{prod,rev} - S_{prod}), \tag{11.10}$$

$$T_0 \Delta S_{irr} = H_{prod,rev} - H_{prod} - T_0\left(S_{prod,rev} - S_{prod}\right). \tag{11.11}$$

Equation 11.11 gives the difference in exergy between the reversible and the irreversible process and hence the loss of exergy, $\Delta E_{loss}$, due to the irreversibility of the process:

$$\Delta E_{loss} = T_0 \Delta S_{irr}. \tag{11.12}$$

This equation is equivalent to Eq. 11.3 representing the law of exergy loss.

The exergy balance is then given as follows::

$$E_+ = E_- + \Delta E_{loss}, \tag{11.13}$$

where the inlet exergy $E_+$ is

$$E_+ = \left\{H_1 - H_{1,0} - T_0\left(S_1 - S_{1,0}\right)\right\} + Q_1\left(1 - \frac{T_0}{T_1}\right), \tag{11.14}$$

and the outlet exergy is

$$E_- = \left\{H_{prod} - H_{prod,0} - T_0\left(S_{prod} - S_{prod,0}\right)\right\} + \left\{H_2 - H_{2,0} - T_0\left(S_2 - S_{2,0}\right)\right\} + Q_0\left(1 - \frac{T_0}{T_0}\right). \tag{11.15}$$

In Eqs. 11.4 and 11.5 the subscript zero indicates the exergy reference state; i.e. the state at the atmospheric temperature and pressure. On the right hand side of Eq. 11.15 the first term is the useful exergy, $E_{prod}$, obtainable from the product, the second term is the rejected exergy, $E_{reject}$, which may be waste or reusable, and the third term is the zero exergy for the discharged heat at the atmospheric temperature: $E_{prod} = H_{prod} - H_{prod,0} - T_0(S_{prod} - S_{prod,0})$; $E_{reject} = H_2 - H_{2,0} - T_0(S_2 - S_{2,0})$. We may call $E_{reject}$ the ***external exergy loss*** in contrast to $\Delta E_{loss}$ which is called the ***internal exergy loss***.

We further define the exergy efficiency, $\eta_E$, as Eq. 11.16:

$$\eta_E = \frac{E_{prod}}{E_+} = 1 - \frac{\Delta E_{loss}}{E_+} - \frac{E_{reject}}{E_+}, \tag{11.16}$$

where $\Delta E_{loss}/E_+$ is the relative internal exergy loss and $\Delta E_{reject}/E_+$ is the relative external exergy loss. Another exergy efficiency called the ***intrinsic exergy efficiency*** has recently been defined by Eq. 11.17 [Ref. 13.], which takes into account the transiting exergy $E_{trasit}$ (the rejected exergy $E_{reject}$):

$$\eta_{intrinsic} = \frac{E_{prod}}{E_+ - E_{trans}} = \frac{E_{prod}}{E_{cons}}, \tag{11.17}$$

where $E_{prod}$ is the exergy for the product and $E_{cons}$ is the exergy actually used for the process itself. The transiting exergy $E_{trans}$ is the part of exergy that enters and traverses the manufacturing process without undergoing any transformation and thereby is not consumed by the process. The intrinsic exergy efficiency usually increases with increasing conversion ratio of the raw material to the product in the process [Ref. 13.].

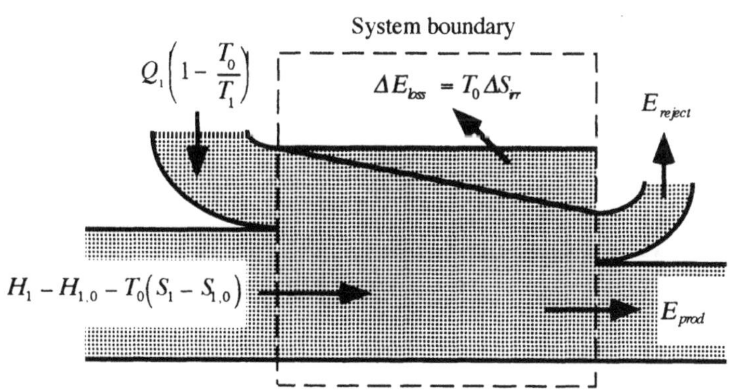


Fig. 11.3. Schematic band diagram of exergy balance for a simple process from a raw material and heat to a product and waste.

Figure 11.3 shows in a schematic way the exergy balance in an irreversible process in the form of a band diagram, where the bandwidth is proportional to the exergy value. For complex physicochemical processes operating in industrial manufacturing plants, the exergy balance diagram is much more complicated than the simple diagram shown in Fig. 11.3, but the exergy analyses in the form of the band diagram is essentially the same no matter how intricate the processes may be. The band diagram of the exergy balance, which is called the Grassmann diagram [Ref. 14.], has widely been used in the exergy analyses of various thermal, mechanical, and physicochemical processes [Refs. 9 and 12.].

### 11. 3. Exergy Vector Diagram.

In order to illustrate the exergy loss in a combination of irreversible physicochemical processes in an open system, Ishida [Ref. 15.] has brought forth the concept of exergy vectors for physical and chemical processes in the form of an enthalpy-exergy diagram. The diagram consists of the ordinate ($y$-axis) indicating the amount of exergy and the abscissa ($x$-axis) indicating the amount of enthalpy of the processes in the system, which we shall call in this book the *exergy vector diagram.* In the following we shall discuss the principle of the exergy vector diagram and its application.

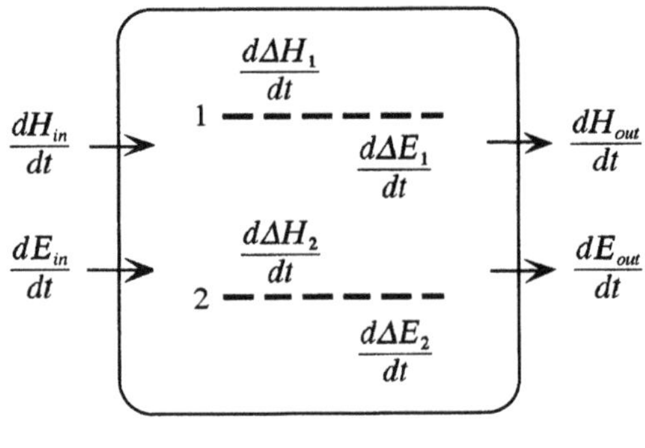


Fig. 11.4. Flows of enthalpy and exergy through an open system in which two processes 1 and 2 are advancing at constant temperature: $H$ = enthalpy flow, $E$ = exergy flow, $\Delta H$ = enthalpy change, $\Delta E$ = exergy change in the system.

Let us consider a simple open system in which two irreversible processes 1 and 2 are occurring in a stationary state as shown in Fig. 11.4. The rates of the enthalpy flow and of the exergy flow between the system and the surroundings are expressed by $dH/dt$ and $dE/dt$, respectively. Further, the rates of the enthalpy change and of the exergy change within the system are expressed by $d\Delta H/dt$ and $d\Delta E/dt$, respectively. The enthalpy and exergy include not only those of substances but also those of work $W$ and heat $Q$. For the transfer of enthalpy Eq. 11.18 holds:

$$H_W = W, \qquad H_Q = Q, \tag{11.18}$$

where $H_W$ and $H_Q$ are the amounts of enthalpy due to work $W$ and heat $Q$, respectively. For the transfer of exergy, in contrast, Eq. 11.19 holds:

$$E_W = W, \qquad E_Q = \left(1 - \frac{T_0}{T}\right) Q, \tag{11.19}$$

where $E_W$ and $E_Q$ are the amounts of exergy due to work $W$ and heat $Q$, respectively.

In accordance with the law of energy conservation (the first law of thermodynamics) Eq. 11.20 is valid for the enthalpy flow and the enthalpy change in the stationary state:

$$\frac{dH_{in}}{dt} - \frac{dH_{out}}{dt} = 0, \qquad \frac{d\Delta H_1}{dt} + \frac{d\Delta H_2}{dt} = 0. \tag{11.20}$$

For the exergy flow and the exergy consumption, however, the law of exergy consumption or entropy creation (the second law of thermodynamics) yields Eq. 11.21:

$$\frac{dE_{in}}{dt} - \frac{dE_{out}}{dt} \geq 0, \qquad \frac{d\Delta E_1}{dt} + \frac{d\Delta E_2}{dt} \leq 0, \tag{11.21}$$

indicating that any irreversible processes consume some amounts of exergy in the system

Figure 11.5 shows a framework of the exergy-enthalpy diagram. The regime of vectors on the left side of the ordinate is for *exothermic* processes where $\Delta H < 0$, while the regime on the right side of it is for *endothermic* processes where $\Delta H > 0$. Furthermore, the regime of vectors on the upper side of the abscissa is for *non-spontaneous* processes where $\Delta E > 0$, while the regime on the lower side of it is for *spontaneous processes* where $\Delta E < 0$.

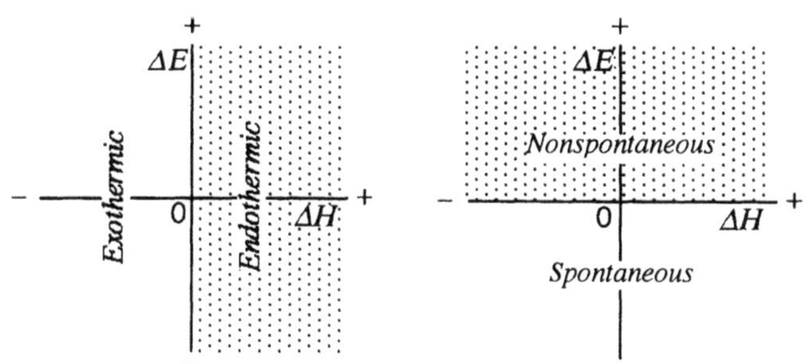


Fig. 11.5. Regimes for exothermic, endothermic, spontaneous, and non-spontaneous processes in exergy-enthalpy diagrams.

Any process that occurs in an open system can be represented on an exergy-enthalpy diagram by a vector starting from the origin of the coordinate axes. A non-spontaneous and endothermic reaction ($\Delta H > 0$, $\Delta E > 0$) such as $Fe_2O_{3\,(solid)} \rightarrow 2Fe_{(liquid)} + 1.5O_{2\,(gas)}$, for example, makes its exergy vector appearing in the regime on the upper and right side of the diagram as shown in Fig. 11.6(a). In this regime, obviously, the enthalpy $\Delta H$ of the vector is the thermal energy that has to be received, and the exergy $\Delta E$ is the exergy that has to be supplied into the system for the reaction to occur.

The reaction, $Fe_2O_{3\,(solid)} \rightarrow 2Fe_{(liquid)} + 1.5O_{2\,(gas)}$, whose exergy change $\Delta E$ has a positive sign, can not spontaneously proceed. But this reaction may be made to proceed, if coupled with an exergy-releasing spontaneous chemical reaction such as $H_{2\,(gas)} + 0.5O_{2\,(gas)} \rightarrow H_2O_{(gas)}$ or $C_{(gas)} + O_{2\,(gas)} \rightarrow CO_{2\,(gas)}$ in the reaction system. This is due to the transfer of exergy from a spontaneous reaction to a non-spontaneous reaction, in which an amount of exergy is released from the spontaneous reaction (an exergy donor) and it is then absorbed by the non-spontaneous reaction (an exergy acceptor). The transferred exergy makes the non-spontaneous reaction proceed. Figure 11.6(b) shows the exergy vectors for a pair of coupling and coupled reactions consisting of a vector for an exergy-releasing process 2 (coupling process) on the lower-left side, a vector for an exergy-absorbing process 1 (coupled process) on the upper-right side, and a composite vector for the combined process of 1 and 2 on the y-axis.

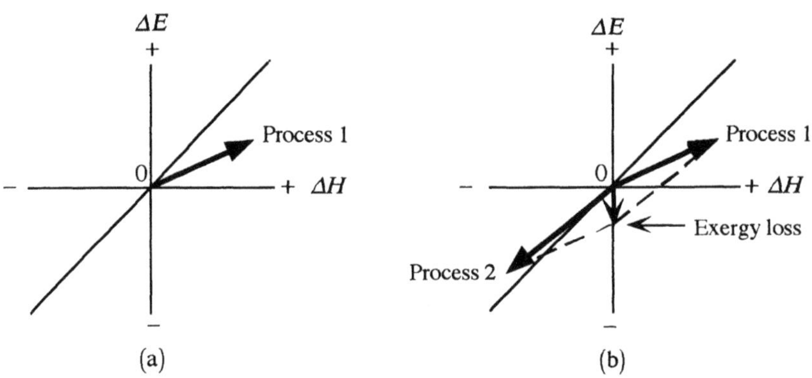


Fig. 11.6. Reaction vectors in exergy-enthalpy diagrams: (a) a non-spontaneous process giving a single reaction vector, (b) coupling and coupled processes leading to a composite reaction vector.

The law of energy conservation makes the *composite vector* lie on the y-axis where $\Delta H = 0$. Furthermore, the law of exergy consumption (entropy creation) for irreversible processes makes the composite vector of the two reactions emerge on the lower side of the y axis where $\Delta E < 0$. The composite vector therefore indicates the amount of exergy consumed (exergy loss) $\Delta E_{loss}$ during the combined process of the coupling and coupled reactions.

If all the processes were occurring reversibly, there would be no consumption of exergy in the system and, hence, the composite vector would vanish at the origin of the coordinates. In order to reduce the exergy consumption, we should therefore make use of such an exergy-releasing reaction so as to reduce the composite vector of the coupling and coupled reactions as close as possible to the origin on the y axis.

In general, a chemical plant manufactures product materials of high exergy levels form raw materials of low exergy levels through exergy-absorbing reactions by using suitable exergy-donating processes. To meet this purpose, we need to use high exergy materials for the coupling reactions and design the manufacturing processes so as to consume the least amount of exergy. The exergy vector diagram shows the losses in exergy and hence serves the purpose of minimizing the consumption of exergy by using the most suitable physical and chemical coupling processes that thermodynamically consume the least amount of exergy in manufacturing the target products.

### 11. 4. Principles in Exergy Vector Diagrams.

The exergy vector of a process in an exergy-enthalpy diagram has a slope $\lambda$ indicating the ratio of the exergy change $\Delta E$ to the enthalpy change $\Delta H$ during the process:

$$\lambda = \frac{\Delta E}{\Delta H} = 1 - \frac{T_0 \Delta S}{\Delta H}, \tag{11.22}$$

where $\Delta S$ is the entropy change in the process: $\Delta E = \Delta H - T_0 \Delta S$.

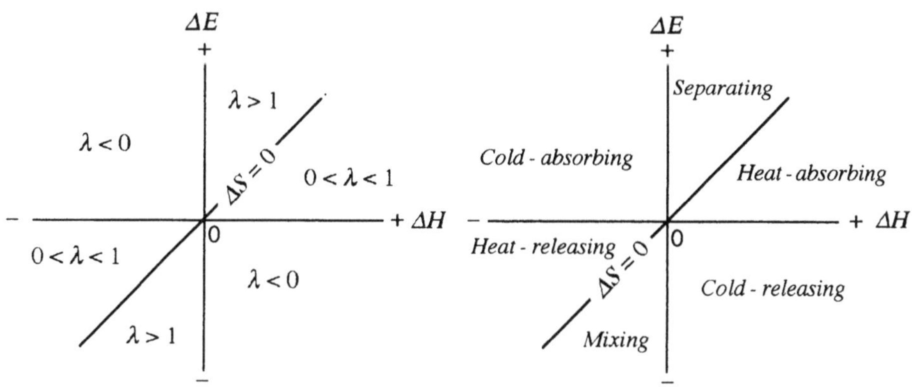


Fig. 11.7. Processes in various regimes in an exergy-enthalpy diagram: $\lambda$ = the energy-availability, line $\Delta S = 0$ corresponds to $\Delta H = \Delta E$.

This ratio $\lambda$ represents the availability of energy in the process. The exergy vector with the energy-availability at $\lambda = 1$ is on the straight line going through the coordinate origin at an angle of 45° as shown in Fig. 11.7. On this line the change $\Delta H$ in enthalpy is equal to the change $\Delta E$ in exergy and hence no entropy change ($\Delta S = 0$) results during the process, indicating that all the energy change is utilized at 100 % efficiency in the form of exergy.

The reversible exchange of work (mechanical, electrical, and other forms of work), in which $\Delta E_W = \Delta H_W$, gives an energy-availability $\lambda_W$ equal to one:

$$\lambda_W = 1, \tag{11.23}$$

which corresponds to a straight line going through the coordinate origin at an angle of 45°. On the other hand, the exchange of heat (thermal energy), in which $\Delta E_Q \neq \Delta H_Q$ even in reversible heat exchange, gives an energy-availability $\lambda_Q$ less than one as follows from Eq. 11.24:

$$\lambda_Q = \left(1 - \frac{T_0}{T}\right), \tag{11.24}$$

indicating that the slope of the exergy vector of thermal processes depends on the temperature $T$ at which heat transfer occurs.

There are three vector regimes in the exergy-enthalpy diagram as described in the following:

(1) The regime in which the energy-availability is in the range of $0 < \lambda < 1$ and in which the changes in entropy $\Delta S$ and in enthalpy $\Delta H$ have the same sign (positive or negative), $1 > T_0\Delta S/\Delta H > 0$: Since an increase or a decrease in both enthalpy and entropy means heat absorption into or heat release from the system, respectively, a non-spontaneous process ($\Delta E > 0$) with positive $T_0\Delta S$ and $\Delta H$ is in the regime of *heat-absorbing processes*, while a spontaneous process ($\Delta E < 0$) with negative $T_0\Delta S$ and $\Delta H$ is in the regime of *heat-releasing processes*.

(2) The regime in which the energy-availability is greater than one $\lambda > 1$ and in which the changes in entropy $\Delta S$ and in enthalpy $\Delta H$ have always the opposite sign (one positive and the other negative), $T_0\Delta S/\Delta H < 0$: The process, in which the enthalpy change is positive $\Delta H > 0$ and the entropy change is negative $\Delta S < 0$, is heat-absorbing, but yet entropy is decreasing despite heat absorption into the system. The decrease in entropy can not result from heat absorption but from separation of the constituent substances in the system; i.e. a decrease in the entropy of mixing. The regime in which $\Delta H > 0$ and $\Delta S < 0$, hence, is for *separating processes*, which absorb heat and decrease entropy of the system. On the other hand, the regime in which $\Delta H < 0$ and $\Delta S > 0$ is for *mixing processes*, which release heat and increase the entropy of the system.

(3) The regime in which the energy-availability is less than zero $\lambda < 0$ and in which the changes in entropy $\Delta S$ and in enthalpy $\Delta H$ have always the same sign (positive sign or negative sign) and $T_0\Delta S$ is larger than $\Delta H$; $T_0\Delta S/\Delta H > 1$: The process, in which the changes

in enthalpy and in entropy are both positive ($\Delta H > 0$ and $\Delta S > 0$) and in which the change in exergy is negative ($\Delta E < 0$), is a spontaneous process releasing negative heat (cold); i.e. a *cold-releasing process*. On the other hand, the process, in which the changes in enthalpy and in entropy are both negative ($\Delta H < 0$ and $\Delta S < 0$) while the change in exergy is positive ($\Delta E > 0$), is a non-spontaneous process absorbing negative heat; i.e. a *cold-absorbing process.*

The term "process" we have used in the foregoing includes physical and chemical changes in general. For example, increasing the temperature of a system from the atmospheric temperature to a high temperature is a heat-absorbing process; decreasing the temperature of a system from a high temperature to the atmospheric temperature is a heat-releasing process; extracting pure oxygen and pure nitrogen from the atmospheric air is a separating process; mixing gaseous oxygen with gaseous nitrogen to produce atmospheric air is a mixing process; cooling a system from the atmospheric temperature down to a low temperature is a cold-absorbing process; and increasing the temperature of a system from a low temperature to the atmospheric temperature is a cold-releasing process.

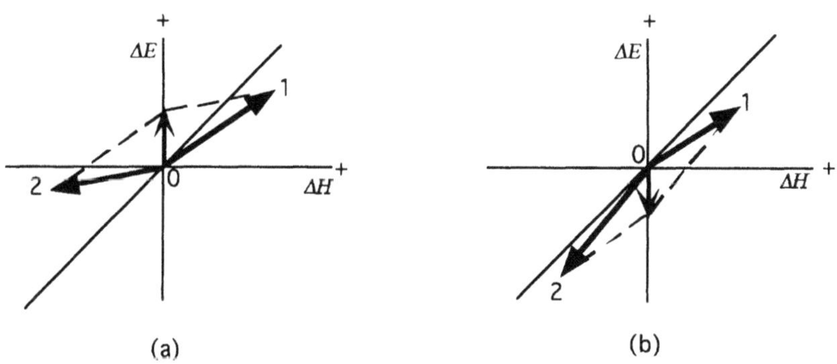


Fig. 11.8. Schematic exergy vector diagrams for exergy transfer between an exergy-absorbing process and an exergy-releasing process: (a) the combined process is not feasible, (b) the combined process is feasible.

## 11.5. Exergy Transfer between Two Processes.

Manufacturing processes in chemical plants involve the exchange of exergy among various physical and chemical processes. Let us now consider a simple case in which an exergy-absorbing process 1 is coupled with an exergy-releasing process 2.

In the case shown in Fig. 11.8(a) where the composite vector, that is the vector sum of the two component vectors of the exergy-absorbing and exergy-releasing processes 1 and 2, points in the direction of exergy increase $(\Delta E > 0)$, the resultant process of the coupled and coupling processes is thermodynamically impossible to occur in the system under consideration. On the other hand, in the case shown in Fig. 11.8(b) where the composite vector points in the direction of exergy decrease $(\Delta E < 0)$, the resultant process is allowed to proceed consuming part of the exergy in the system.

The feasibility of the resultant process depends on the slopes $\lambda_1$ and $\lambda_2$ of the two vectors of processes 1 and 2; $\lambda$ is the energy-availability discussed in section 11.4. In the case where the slope $\lambda_1$ of the vector of *exergy-absorbing process* 1 is greater than the slope $\lambda_2$ of the vector of *exergy-releasing process* 2, no transfer of exergy can occur between the two processes and hence the resultant process is not feasible; whereas, in the reverse case where $\lambda_1$ is smaller than $\lambda_2$ the transfer of exergy is allowed to occur from process 2 to process 1 and then the resultant process may occur spontaneously. The exergy transfer therefore requires that the energy-availability $\lambda_{donor}$ of the exergy-releasing process should be greater than the energy-availability $\lambda_{acceptor}$ of the exergy-absorbing process:

$$\lambda_{donor} > \lambda_{acceptor}. \tag{11.25}$$

This is the criterion for the transfer of exergy between the donor and the acceptor process.

## 11. 6. Exergy Vectors of Heating-cooling Processes.

Heating and cooling are commonly occurring processes in chemical plants. Let us consider a thermal process in which an amount of heat, $Q$, is transferred into or out of a system at a constant temperature $T$. Equation 10.24 gives us the energy-availability $\lambda_Q$ in heating and cooling processes as shown in Eq. 11.26:

$$\lambda_Q = 1 - \frac{T_0}{T}, \tag{11.26}$$

where $T > T_0$ for heat-absorbing processes, while $T < T_0$ for cold-absorbing (heat-releasing) processes. The energy-availability $\lambda_Q$ equals the slope of the exergy vector of the process.

The exergy vectors of heating-cooling processes, as shown in Fig. 11.9, are exemplified by a vector with the energy-availability of $\lambda_Q = 1$ (vector slope 45°, $\Delta S = 0$, $\Delta H = \Delta E$) for heat transfer at apparently unlimited high temperature $T = \infty$ K (e.g. electrical heating); by a vector with the slope of $\lambda_Q = 0.5$ for heat transfer at $T = 596$ K in the heat-absorbing regime on the upper right hand side and in the heat-releasing regime on the lower left hand side of the diagram; by a vector on the abscissa (enthalpy axis) with the slope of $\lambda_Q = 0$ for heat transfer at the atmospheric temperature $T = T_0$ where heat possesses no exergy; by a vector with the slope of $\lambda_Q = -0.5$ for cold transfer (inverse heat transfer) at $T = 199$ K in the cold-absorbing regime on the upper left side and in the cold-releasing regime on the lower right hand side of the diagram; by a vector with the slope of $\lambda_Q = -1.5$ for cold transfer at $T = 119$ K in the cold-absorbing and -releasing regimes; and by a vector on the ordinate (exergy axis) with the slope of $\lambda_Q = -\infty$ for cold transfer at the temperature $T = 0$ K. In the heating and cooling processes the energy-availability $\lambda_Q$, thus, increases with increasing temperature from $\lambda_Q = -\infty$ at the temperature zero $T = 0$ K through $\lambda_Q = 0$ at the atmospheric temperature $T = T_0$ up to $\lambda_Q = 1$ at apparently unlimited high temperature $T \to \infty$.

We notice in Fig. 11.9 that no vector of thermal processes can occupy the separating and mixing regimes in the exergy vector diagram.

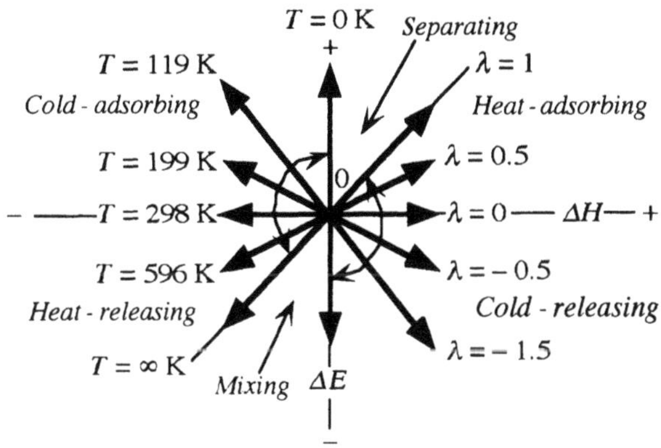


Fig. 11. 9. Exergy vectors for heating and cooling processes at constant temperature.

Table 11.1. Energy availability of heating-cooling processes at constant temperature.

| Temperature K | Energy Availability $\lambda_Q$ |
|---|---|
| ∞ | + 1.0 |
| 2980 | + 0.9 |
| 1490 | 0.8 |
| 596 | + 0.5 |
| 298 | 0.0 |
| 199 | - 0.5 |
| 149 | - 1.0 |
| 119 | - 1.5 |
| 0 | – ∞ |

Table 11.1 shows the energy availability $\lambda_Q$ of heating and cooling processes as a function of the temperature at which the heat transfer occurs. The positive sign of $\lambda_Q$ denotes the transfer of heat and the negative sign of $\lambda_Q$ denotes the transfer of cold. It is a matter of course that the enthalpy component $\Delta H$ of the vector for heating and cooling processes is the

amount of heat $Q$ received by or released from the system. The exergy component of the vector represents the absorption or release of available energy during the heat or cold transfer at constant temperature.

## 11. 7. Exergy Vectors of Compression-expansion Processes.

We have examined in section 10.3 the exergy associated with a mechanical work of compression and expansion of gases. The process of *isentropic compression-expansion* (*adiabatic compression-expansion*) is represented by an exergy vector lying on the line of $\lambda_W = 1$ that passes through the origin of the coordinates with the slope of $45^\circ$ $(\Delta S = 0)$ as shown in Fig. 11.10(a); the vector points in the positive direction for compression and in the negative direction for expansion. The enthalpy component $\Delta H_W$ of the exergy vector of isentropic compression-expansion represents the amount of work received ($\Delta H_W > 0$) by or released ($\Delta H_W < 0$) from the system.

In contrast to isentropy, the process of *isothermal compression-expansion*, which is accompanied by heat release or heat absorption, is represented by an exergy vector with the slope of $\lambda_W < 1$ in the regimes of heat absorption and heat release as shown in Fig. 11.10(a).

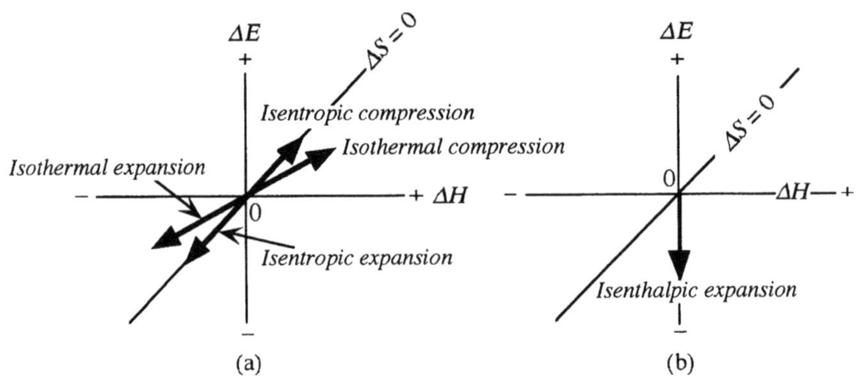


Fig. 11.10. Exergy vectors for gaseous compression and expansion processes.

There is another mode of gaseous expansion called the Joule-Thomson expansion, in which the change in gas volume occurs at constant enthalpy $\Delta H_W = 0$ without any change in energy. The vector of the *isenthalpic expansion* then stands perpendicular to the abscissa on the ordinate and points in the negative direction (exergy consumption) as is shown in Fig. 11.10(b).

**11. 8. Exergy Vectors of Separating-mixing Processes.**

We have discussed in section 10.6 exergy changes associated with the processes of separating and mixing the constituent substances in a mixture. Separating and mixing of substances in a perfect mixture, such as an ideal gas mixture and a perfect solution, are accompanied by no change in enthalpy $(\Delta H = 0)$ so that their exergy vectors lie on the ordinate (y-axis) pointing in the positive direction $(\Delta E > 0)$ for the separating process and in the negative direction $(\Delta E < 0)$ for the mixing process as shown in Fig. 11.11(a).

In a non-ideal solution, however, the separating process absorbs an amount of heat into the solution and the mixing process releases an amount of heat out of the solution. Their exergy vectors consequently emerge in the separating regime $(\Delta H > 0)$ and in the mixing regime $(\Delta H < 0)$, respectively, as shown in Fig. 11.11(b).

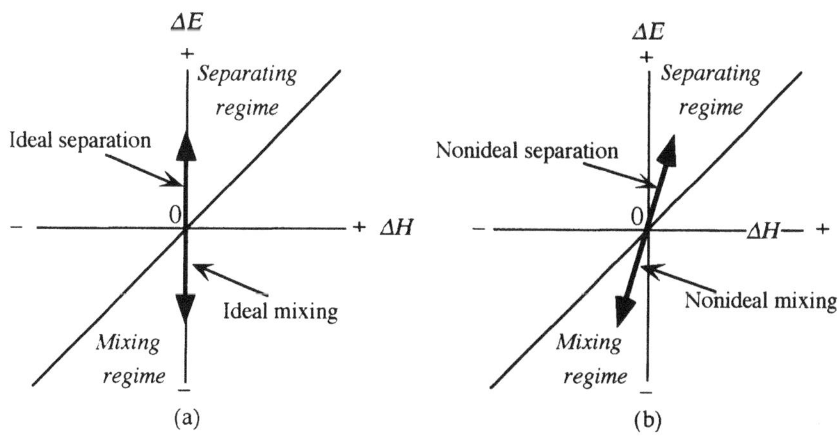


Fig. 11.11. Exergy vectors for separating and mixing processes.

**11. 9. Exergy Vectors of Chemical Reactions.**

We have discussed the exergy of chemical substances in chapter 10. The exergy associated with a chemical reaction is given by the difference in exergy between the reactants and the products and hence by the difference in free enthalpy between the reactants and the products, while the enthalpy of the chemical reaction is given by the heat of the reaction.

In general, the exergy vector of an exergy-absorbing reaction $(\Delta E > 0)$ stands in the heat-absorbing regime or in the separating regime where the enthalpy of the reaction is positive $(\Delta H > 0$, endothemic reaction$)$. Figure 11.12 shows, as examples, the exergy vectors for the following exergy-absorbing reactions at the standard temperature (298 K) and pressure (101.3 kPa):

$$H_2O_{(gas)} \rightarrow H_{2(gas)} + 0.5\, O_{2(gas)}\,,$$

$$FeO_{(solid)} \rightarrow Fe_{(solid)} + 0.5\, O_{2(gas)}\,,$$

$$C_{(solid)} + H_2O_{(gas)} \rightarrow CO_{(gas)} + H_{2(gas)}\,.$$

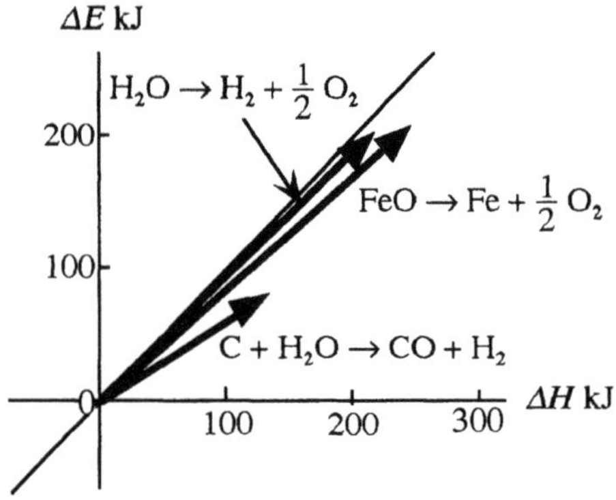


Fig. 11.12. Exergy vectors for exergy-absorbing reactions [Ref. 15.].

On the other hand, the exergy vector of an exergy-releasing reaction $(\Delta E < 0)$ stands in the heat-releasing regime or in the mixing regime where the enthalpy of the reaction is negative $(\Delta H < 0$, exothemic reaction$)$. Figure 11.13 shows, for examples, the exergy vectors of the following exergy-releasing reactions at the standard temperature (298 K) and pressure (101.3 kPa):

$$H_{2(gas)} + 0.5\, O_{2(gas)} \rightarrow H_2O_{(gas)}\,,$$

$$C_{(solid)} + 0.5\, O_{2(gas)} \rightarrow CO_{(gas)}\,,$$

$$C_{(solid)} + O_{2(gas)} \rightarrow CO_{2(gas)}\,,$$

$$CO_{(gas)} + 0.5\, O_{2(gas)} \rightarrow CO_{2(gas)}\,,$$

$$CO_{(gas)} + 2\, H_{2(gas)} \rightarrow CH_3OH_{(gas)}\,.$$

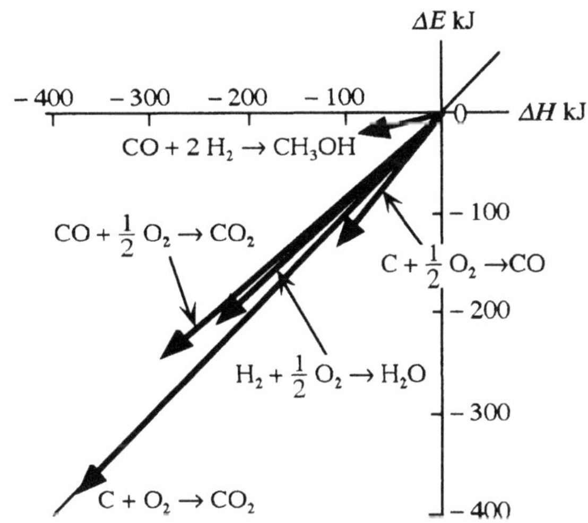


Fig. 11.13. Exergy vector for exergy-releasing reactions [Ref. 15.].

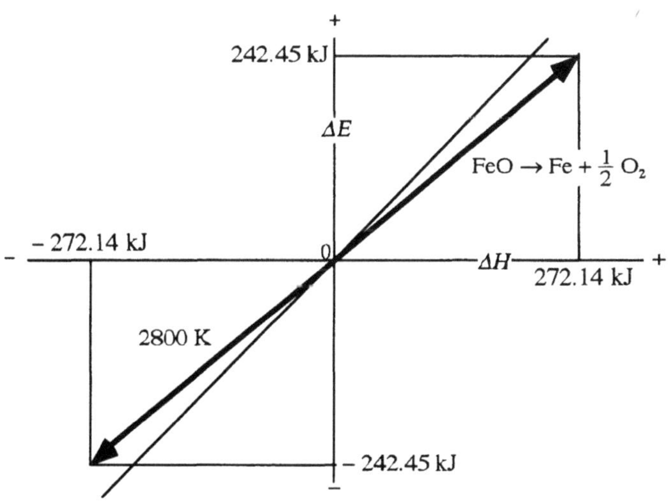


Fig. 11.14. Exergy vector diagram for an iron oxide reduction, $FeO \rightarrow Fe + 0.5O_2$, coupled with a heating process at 2800 K [Ref. 15.].

## 11. 10. Exergy Transfer in Chemical Reactions.

We now consider, as an example, the non-spontaneous reaction of iron oxide reduction, $FeO_{(solid)} \rightarrow Fe_{(solid)} + 0.5\ O_{2\,(gas)}$, in which the heat of reaction is $\Delta H = 272.14$ kJ (endothermic) and the exergy of reaction is $\Delta E = 242.45$ kJ (exergy-absorbing) at the standard temperature (298 K) and pressure (101.3 kPa). In Fig. 11.14 the exergy vector of this reaction is shown whose energy-availability (the slope of the vector ) is $\lambda = 0.895$. As mentioned in the foregoing (section 11.5), in order to make iron oxide reduced, we need to couple this reduction reaction with an exergy-releasing process whose energy-availability is greater than 0.895.

To a first approximation, recalling the heat of chemical reactions almost independent of temperature, we assume that $\Delta H$ and $\Delta E$ of the iron oxide reduction do not change much with temperature in the temperature range considered. If we make use of a thermal process to supply an adequate amount of exergy toward the reduction of FeO, a heating process is needed whose energy-availability is greater than $\lambda = 0.895$, namely heating at a temperature higher than 2800 K, with a supply of thermal energy more than $\Delta H_{therm+} = 272.14$ kJ for one mole of the iron oxide as shown in Fig. 11.14: $(T - T_0)/T = \lambda = 0.895$ gives us $T \approx 2800$ K.

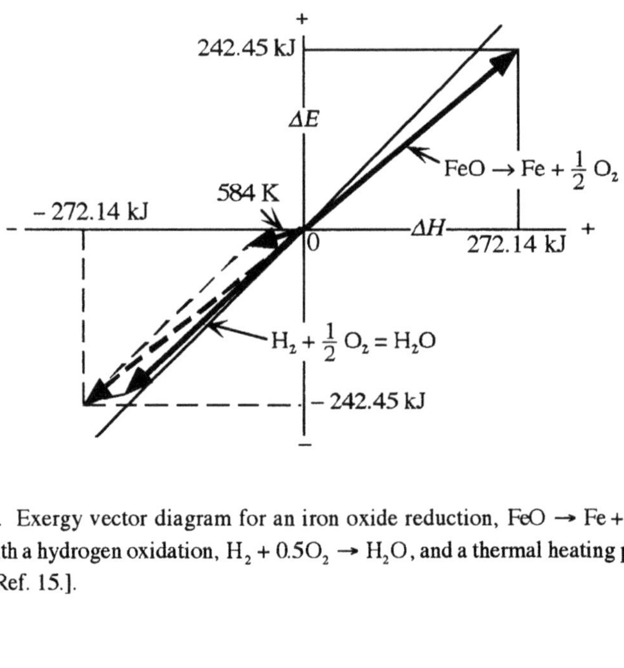


Fig. 11.15. Exergy vector diagram for an iron oxide reduction, $FeO \rightarrow Fe + 0.5O_2$, coupled with a hydrogen oxidation, $H_2 + 0.5O_2 \rightarrow H_2O$, and a thermal heating process at 584 K [Ref. 15.].

As an alternative to the thermal exergy for reducing FeO at 2800 K , we may use as an exergy donor the oxidation of hydrogen gas, $H_{2(gas)} + 0.5\,O_{2(gas)} \rightarrow H_2O_{(gas)}$. This reaction of hydrogen oxidation provides us with an amount of enthalpy $\Delta H = -241.83$ kJ and an amount of exergy $\Delta E = \Delta E = -228.59$ kJ giving the energy availability (the slope of the exergy vector) at $\lambda = 0.945$. This slope $\lambda = 0.945$ of the exergy vector for the hydrogen oxidation is greater than the slope $\lambda = 0.895$ of the exergy vector for the iron oxide reduction so that the transfer of exergy from the former to the latter is thermodynamically possible. Hydrogen oxidation with $\Delta H = -241.83$ kJ, however, is unable to donate the enough energy $\Delta H =$

272.14 kJ required for the reduction of the iron oxide. To make up for a deficit in energy $\Delta H_Q = -30.31$ kJ for one mole of FeO, an additional thermal process is needed such as a heating process at a temperature 584 K to be able to reduce FeO as estimated by the exergy vector analysis shown in Fig. 11.15. The overall reaction is then given by $FeO_{(solid)} + H_{2\,(gas)} \rightarrow Fe_{(solid)} + H_2O_{(gas)}$. We note that, employing the oxidation of hydrogen as an exergy-donating process, we may reduce the temperature of the thermal energy required for reducing FeO from 2800 K to 584 K, which is the theoretical lowest temperature for the reaction to proceed: We need in reality the heating process at least higher than 800 K for the hydrogen reduction of iron oxides to occur.

We may also make use of the reaction of carbon monoxide formation, $C_{(solid)} + (1/2)\,O_{2(gas)} \rightarrow CO_{(gas)}$. This reaction provides us with an amount of enthalpy $\Delta H = -110.52$ kJ and an amount of exergy $\Delta E = -137.27$ kJ, giving the energy availability (the slope of the exergy vector) at $\lambda = 1.242$. This slope $\lambda = 1.242$ of the exergy vector for the carbon monoxide formation is greater than the slope $\lambda = 0.895$ of the exergy vector for the iron oxide reduction so that the transfer of exergy from the former to the latter is thermodynamically possible. Carbon monoxide formation with $\Delta H = -110.52$ kJ, however, is unable to donate the enough amount of energy $\Delta H = 272.14$ kJ required for the reduction of FeO. To supply a deficit amount of energy $\Delta H_Q = -161.62$ kJ for the iron oxide reduction, an additional thermal process such as a heating process at temperatures higher than 869 K is needed as estimated by the exergy vector analyses shown in Fig. 11.16. The overall reaction then is expressed by: $FeO_{(solid)} + C_{(solid)} \rightarrow Fe_{(solid)} + CO_{(gas)}$.

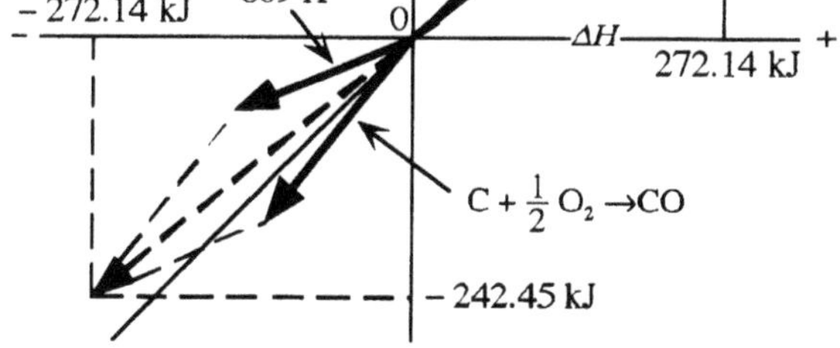


Fig. 11.16. Exergy vector diagram for an iron oxide reduction, $FeO \rightarrow Fe + 0.5O_2$, coupled with a carbon oxidation, $C + 0.5O_2 \rightarrow CO$, and a thermal heating process at 869 K [Ref. 15.].

The oxidation of carbon to carbon monoxide, as shown in Figs. 11.13 and 11.16, has an energy availability $\lambda$ greater than one, and accordingly the exergy vector takes its position in the regime of mixing, indicating that the reaction includes a mixing process which donates as a whole the greater amount of exergy than the heat of the reaction. We note that the greater the energy availability and hence the greater the slope of the exergy vector of an exergy-donating process is, then the more effective the exergy transfer becomes.

## 11. 11. Exergy Vector Diagrams of Methanol Synthesis.

We examine, as an example, the exergy vector diagram for methanol synthesis to estimate the minimum exergy loss thermodynamically required for the synthesis reaction of methanol from methane [Ref. 16.]. First, we consider a direct (single step) synthesis of methanol from methane through a coupled-and-coupling reaction consisting of the oxidation of methane (objective reaction) and the dissociation of water molecule (coupled reaction) shown, respectively, as follows:

$$CH_4 + 0.5O_2 \rightarrow CH_3OH, \quad \Delta H = -126.38 \text{ kJ/mol}, \quad \Delta E = -110.89 \text{ kJ/mol},$$

$$H_2O_{liq} \rightarrow H_2 + 0.5O_2, \quad \Delta H = 285.99 \text{ kJ/mol}, \quad \Delta E = 237.30 \text{ kJ/mol},$$

where $\Delta H$ and $\Delta E$ are the enthalpy change and the exergy change of the reactions, respectively.

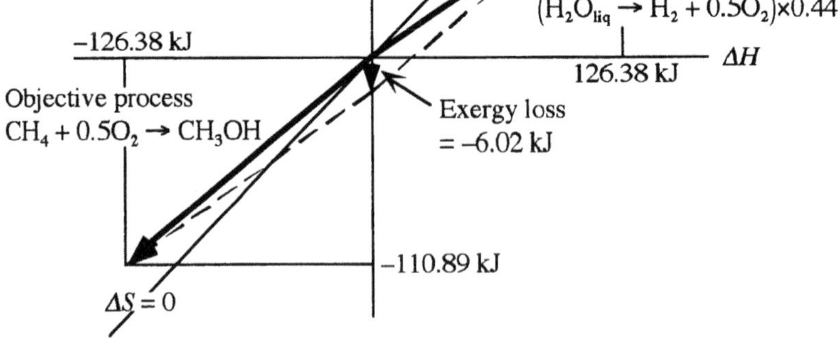


Fig. 11.17. Exergy diagram for a direct synthesis of methanol showing the theoretical minimum exergy consumption [Ref. 16.].

Based on the law of energy conservation, the enthalpy changes of the two reactions must be balanced in the stationary state so that the stoichiometrical ratio in energy of the two reactions is 0.44: The enthalpy of methane oxidation $CH_4 + 0.5O_2 \rightarrow CH_3OH$ is 0.44 times as much as the enthalpy of water dissociation $H_2O_{liq} \rightarrow H_2 + 0.5O_2$. Furthermore, the law of exergy decrease predicts that the composite exergy vector of the two reactions must be on the exergy axis (ordinate) pointing to the negative direction. The exergy vector diagram thus obtained is shown in Fig. 11.17. We thus estimate the theoretical minimum exergy loss $\Delta E_{theo,\,loss}$ required for the direct synthesis of methanol is $\Delta E_{theo,\,loss} = 6.02$ kJ/mol - $CH_3OH$:

$$\text{Direct methanol synthesis,} \quad \Delta E_{theo,\,loss} = 6.02 \text{ kJ/mol - CH}_3\text{OH}(0.19 \text{ GJ/t - methanol}).$$

Methanol manufacturing processes in current use have been reported to consume forty four times as much exergy as the theoretical minimum exergy loss estimated above for the direct methanol synthesis [Ref. 16.].

Methanol can also be produced through a two-step process comprising of steam reforming of methane and methanol synthesis from carbon monoxide and hydrogen. The first step of steam reforming of methane consists of the following two reactions:

$$CH_4 + H_2O \rightarrow CO + 3H_2, \quad \Delta H = 250.28 \text{ kJ/mol}, \quad \Delta E = 150.95 \text{ kJ/mol},$$

$$H_2 + 0.5O_2 \rightarrow H_2O_{gas}, \quad \Delta H = -241.95 \text{ kJ/mol}, \quad \Delta E = -228.72 \text{ kJ/mol}.$$

The enthalpy of methane reforming, $CH_4 + H_2O \rightarrow CO + 3H_2$, is balanced against 0.88 times the enthalpy of hydrogen oxidation, $H_2 + 0.5O_2 \rightarrow H_2O_{gas}$. In the same way as is used for the single step methanol synthesis, we obtain the theoretical minimum exergy loss required for the steam reforming to be $\Delta E^1_{theo,\,loss} = 52.60$ kJ/mol - $CH_3OH$ as shown in Fig. 11.8.

The second step of methanol synthesis from carbon monoxide consists of the following two reactions:

$$CO + 2H_2 \rightarrow CH_3OH, \quad \Delta H = -90.67 \text{ kJ/mol}, \quad \Delta E = -24.53 \text{ kJ/mol},$$

$$H_2O_{liq} \rightarrow H_2O_{gas}, \quad \Delta H = 44.04 \text{ kJ/mol}, \quad \Delta E = 8.58 \text{ kJ/mol}.$$

The enthalpy of the hydrogenation of carbon monoxide, $CO + 2H_2 \rightarrow CH_3OH$, is balanced against 2.06 times the enthalpy of water evaporation, $H_2O_{liq} \rightarrow H_2O_{gas}$. For this step of methanol synthesis from carbon monoxide the theoretical minimum exergy loss is $\Delta E^2_{theo,\,loss} =$ 6.86 kJ/mol - $CH_3OH$ as shown in Fig. 11.8. We hence obtain $\Delta E_{theo,\,loss} =$ 59.46 kJ/mol - $CH_3OH$ as a whole for the theoretical minimum exergy loss thermodynamically required for producing methanol by means of the two-step synthesis from methane:

$$\text{Two-step methanol synthesis,} \quad \Delta E_{theo,\,loss} = 59.46 \text{ kJ/mol - CH}_3\text{OH}(1.86 \text{ GJ/t - methanol}).$$

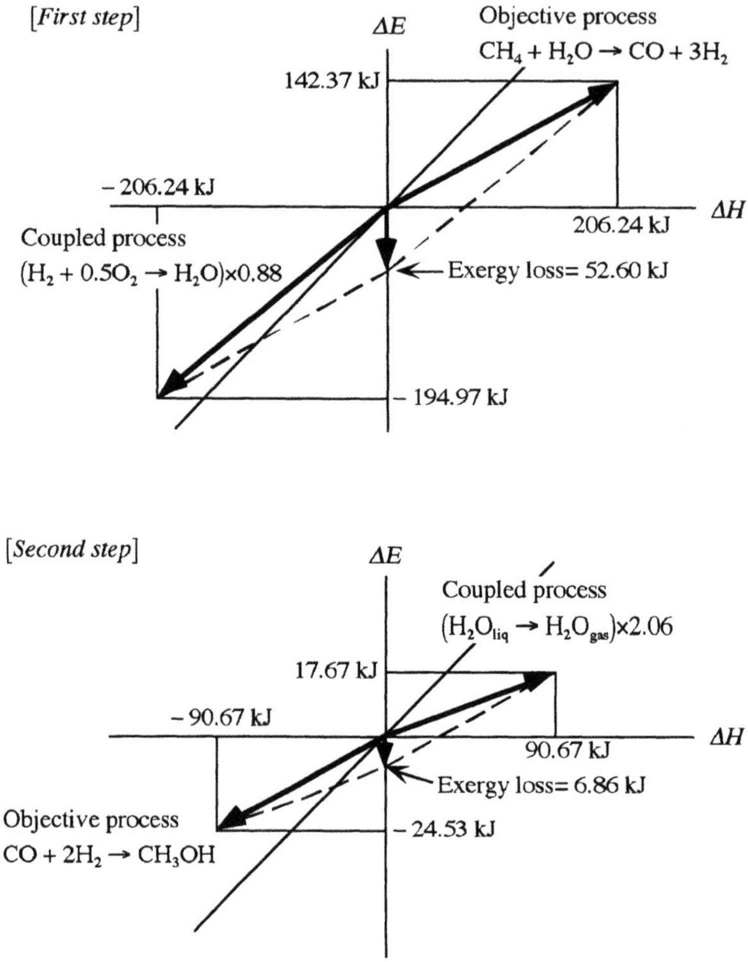


Fig. 11.18. Schematic exergy diagram for theoretical minimum exergy consumption in a two-step synthesis of methanol [Ref. 16.].

The foregoing estimation of the theoretical minimum exergy loss $\Delta E_{theo,\,loss}$ shows that the value of $\Delta E_{theo,\,loss}$ of the direct methanol synthesis is one-tenth that of the two-step methanol synthesis. It then follows that the direct synthesis of methanol is advantageous over the two-step synthesis in the efficient use of exergy.

**11. 12. Exergy Vectors for Electrochemical Reactions.**

We now examine the exergy vectors of electrochemical reactions for water electrolysis and hydrogen-oxygen fuel cells at the atmospheric temperature. The electrochemical reaction of water electrolysis is expressed as follows:

$$H_2O_{liq} \rightarrow H_{2,\,gas} + 0.5\,O_{2,\,gas}, \quad \Delta H = 285.83\ \text{kJ/mol}, \quad \Delta E = 237.18\ \text{kJ/mol};$$

and the reverse of this reaction is the hydrogen-oxygen fuel cell reaction:

$$H_{2,gas} + 0.5\,O_{2,gas} \rightarrow H_2O_{liq}, \quad \Delta H = -285.83\ \text{kJ/mol}, \quad \Delta E = -237.18\ \text{kJ/mol};$$

where $\Delta H$ and $\Delta E$ are the enthalpy change and the exergy change of the reactions at the standard state (the atmospheric temperature and pressure), respectively; $\Delta E$ being equal to the free enthalpy change $\Delta G$ of the reaction.

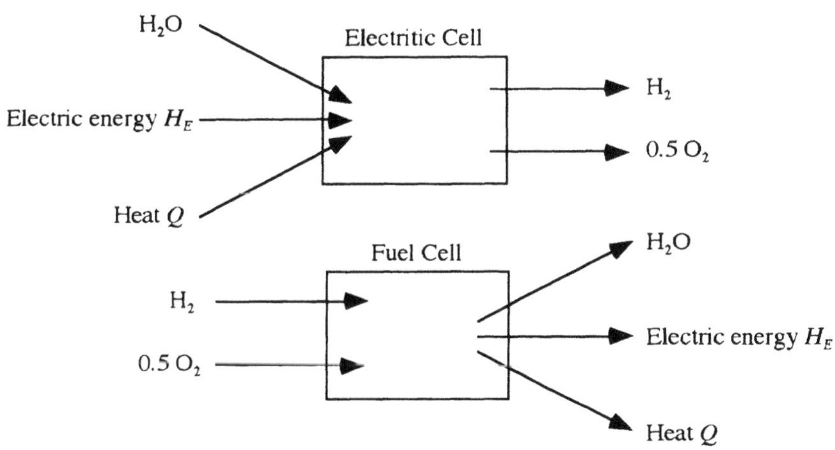


Fig. 11.19. Schematic processes of water electrolysis cells and hydrogen-oxygen fuel cells at room temperature.

Figure 11.19 shows the processes that occur in the electrolytic cell and in the fuel cell. Electric energy contains no entropy when it provides for the cell or extracts from the cell an amount of electrical work, and hence its energy-availability $\lambda_{WE}$ equals one; $\lambda_{WE} = 1$. The exergy vector of electric energy consequently is located on the straight line going through the coordinate origin at an angle of $45^\circ$. On the other hand, the heat transfer at the atmospheric temperature, if occurring reversibly, produces no exergy change and its exergy vector therefore appears on the abscissa (enthalpy axis).

Figure 11.20 shows the exergy vector diagram for the reversible process of water electrolysis. The exergy vector of electric energy $H_{WE}$ supplied at the energy-availability $\lambda_{WE}$ equal to one and the exergy vector of reversible heat transfer $Q$ supplied at the energy-availability $\lambda_Q$ equal to zero are combined together to make a composite exergy vector of the decomposition of water molecules into molecular hydrogen and oxygen gases.

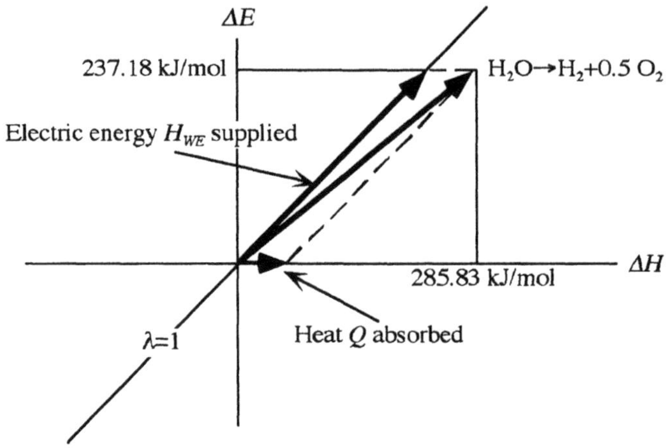


Fig. 11.20. Exergy vector diagram of water electrolysis reaction at room temperature.

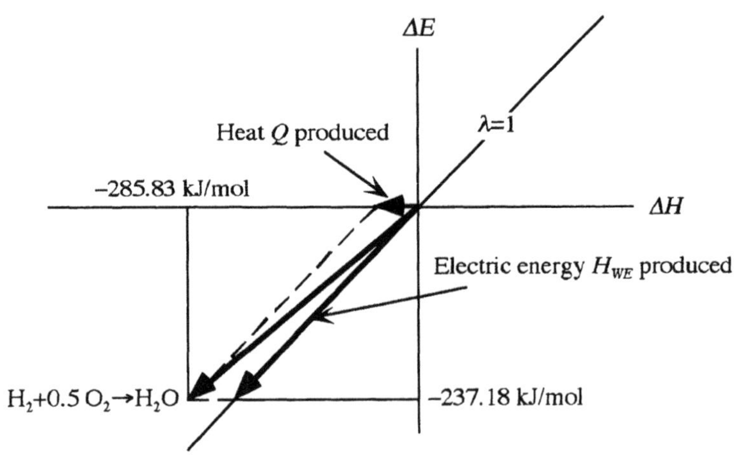


Fig. 11.21. Exergy vector diagram of hydrogen-oxygen fuel cell reaction at room temperature.

The exergy vector diagram for the reversible process of a hydrogen-oxygen fuel cell is shown in Fig. 11.21, in which the exergy vector of the formation of water molecules from gaseous hydrogen and oxygen molecules electrochemically decomposes into both an exergy vector of electric energy $H_{WE}$ produced and an exergy vector of reversible heat transfer $Q$ released from the cell at the atmospheric temperature. When the heat transfer $Q$ occurs irreversibly at temperature higher than the atmospheric temperature, the exergy vector of heat transfer $Q$ deviates from the abscissa, and hence the exergy vector of electric energy $H_{WE}$ produced by the cell is reduced causing an internal exergy loss in the cell.

# LIST OF SYMBOLS

| | |
|---|---|
| $A$ | Affinity of irreversible process (reaction). |
| $A^*$ | Unitary affinity of irreversible process. |
| $A^M$ | Affinity of mixing in irreversible process. |
| $A^0$ | Standard affinity of irreversible process. |
| $a_{T,p}$ | Differential of affinity with respect to extent of reaction at constant temperature and pressure $(\partial A/\partial \xi)_{T,p}$. |
| $a_i$ | activity of chemical substance $i$. |
| $\boldsymbol{a}_i$ | absolute activity of chemical substance $i$. |
| $C_{p,\xi}$ | heat capacity at constant pressure and composition. |
| $C_{V,\xi}$ | heat capacity at constant volume and composition. |
| $c_i$ | molar concentration of substance $i$ in molarity scale. |
| $c_{p,i}$ | partial molar heat capacity of $i$ at constant pressure. |
| $c_p^E$ | partial molar excess heat capacity at constant pressure. |
| $E$ | electrode potential. |
| $E_{eq}$ | equilibrium electrode potential. |
| $E_{\mathrm{EMF}}$ | electromotive force of electrochemical cell. |
| $E^0$ | standard equilibrium electrode potential. |
| e | electron. |
| $\mathrm{e}_{(\mathrm{REDOX})}$ | redox electron. |
| $F$ | free internal energy (Helmholtz energy), Faraday constant. |
| $f_i$ | fugacity of substance $i$. |
| G | free enthalpy (Gibbs energy). |
| $G^E$ | excess free enthalpy. |
| $g_i$ | partial molar free enthalpy of substance $i$. |
| $g^E$ | partial molar excess free enthalpy. |
| $g_{\mathrm{mean}}$ | average partial molar free enthalpy. |
| $H$ | enthalpy. |
| $H^E$ | excess enthalpy. |
| $h_i$ | partial molar enthalpy of substance $i$. |
| $h_{T,p}$ | differential of heat of reaction at constant pressure. |

| | |
|---|---|
| $h_{T,\xi}$ | latent heat of pressure change at constant temperature and composition. |
| $h_i^*$ | unitary partial molar enthalpy of substance $i$. |
| $h_i^M$ | partial molar enthalpy of mixing for substance $i$. |
| $h^E$ | partial molar excess enthalpy. |
| $K$ | equilibrium constant of reaction. |
| $l_{T,\xi}$ | latent heat of volume change at constant temperature and composition |
| $M_i$ | molecular weight. |
| $m$ | mass. |
| $m_i$ | concentration of substance $i$ in molality scale. |
| $N$ | number of particles. |
| $n_i$ | number of moles of substance $i$. |
| $p$ | pressure. |
| $Q$ | heat. |
| $Q_{av}$ | average heat of reaction. |
| $Q_{irr}$ | uncompensated heat created in irreversible process. |
| $Q_{T,V}$ | heat of reaction at constant volume. |
| $Q_{T,p}$ | heat of reaction at constant pressure. |
| $q$ | differential heat of reaction. |
| $S$ | entropy. |
| $S_{irr}$ | entropy created in irreversible process. |
| $S^M$ | entropy of mixing. |
| $S^E$ | excess entropy. |
| $s$ | partial molar entropy. |
| $s_i^*$ | unitary partial molar entropy of substance $i$. |
| $s_i^0$ | standard molar entropy of pure substance $i$. |
| $s_i^M$ | partial molar entropy of mixing for substance $i$. |
| $s^E$ | partial molar excess entropy. |
| $T$ | absolute temperature. |
| $U$ | internal energy. |
| $u_{T,\xi}$ | differential heat of reaction at constant volume. |
| $V$ | volume. |
| v | rate of reaction. |
| $v_i$ | partial molar volume of substance $i$. |
| $v_i^*$ | unitary partial molar volume of substance $i$. |
| $v_i^M$ | partial molar volume of mixing for substance $i$. |
| $v^E$ | partial molar excess volume. |
| $v_i^{pure}$ | molar volume of pure substance $i$. |
| $W$ | work done. |
| $X_i$ | driving force of irreversible process $i$. |
| $x_i$ | molar fraction of substance $i$. |
| $z_i$ | valence of ion $i$. |

| | |
|---|---|
| $\alpha$ | coefficient of thermal expansion. |
| $\alpha_e$ | real potential of electron (minus work function). |
| $\alpha_i$ | real potential of charged particle $i$. |
| $\gamma_i$ | activity coefficient of substance $i$. |
| $\gamma_\pm$ | mean activity coefficient of ions (cation and anion). |
| $E$ | exergy. |
| $\varepsilon$ | partial molar exergy. |
| $\varepsilon_i$ | energy of particle $i$, exergy of particle $i$. |
| $\eta_e$ | electrochemical potential of electron. |
| $\eta_i$ | electrochemical potential of charged particle $i$. |
| $\kappa$ | coefficient of compressibility. |
| $\lambda$ | availability rate of energy. |
| $\mu_i$ | chemical potential of substance $i$. |
| $\mu_\pm$ | mean chemical potential of a pair of cation and anion. |
| $\mu_i^*$ | unitary chemical potential of substance $i$. |
| $\mu_i^M$ | chemical potential of mixing for substance $i$. |
| $\mu_i^{\text{pure}}$ | chemical potential of pure substance $i$. |
| $\mu_i^0$ | standard chemical potential of chemical substance $i$.. |
| $\nu_i$ | stoichiometrical coefficient of substance $i$ in reaction. |
| $\xi$ | extent of irreversible process (reaction). |
| $\pi$ | osmotic pressure. |
| $\phi$ | osmotic coefficient. |
| $\phi$ | inner potential. |
| $\chi$ | surface potential (surface potential difference). |
| $\psi$ | outer potential. |
| $\Omega$ | number of distinct microscopic states of system. |

# REFERENCES


(1) I. Prigogine and R. Defay (Translated by D. H. Everett), "Chemical Thermodynamics", Longmans, Green and Co. Ltd., London (1962).

(2) I. Prigogine, "Thermodynamics of Irreversible Processes", Interscience Publishers, New York, London, Sydney (1967).

(3) R. W. Gurney, "Ionic Processes in Solution", pp. 80-112, McGraw Hill Book Co. Inc., New York (1953).

(4) N. Sato, "Electrode Potentials", Materia Japan, Vol. 33, p. 1390, Japan Institute of Metals, Sendai (1994).

(5) N. Sato, "Electrochemistry of Metal and Semiconductor Electrodes", Elsevier, Amsterdam (1998).

(6) F. Bosnjakovic, " Technical Thermodynamics", Holt, Rinehart & Winston, New York (1956).

(7) T. Nobusawa, "Introduction to Exergy", Ohm Pub., Tokyo (1984).

(8) Z. Rant, Forsch. Ing. Wes., Vol. 22, pp. 36-37(1956).

(9) J. Szargut, D. R. Morris, and F. R. Steward, "Exergy Analysis of Thermal, Chemical, and Metallurgical Processes", Hemisphere Publishing Corporation, New York, London (1988).

(10) L. Riekert, Chemical Eng. Science, Vol. 29, pp. 1613-1620(1974).

(11) D. R. Morris and F. R. Steward, "Exergy Analysis of a Chemical Metallurgical Process", Metallurgical Transactions, Vol. 15B, pp. 645-654(1984).

(12) T. Akiyama and J. Yagi, "Exergy", Bulletin of The Iron and Steel Institute of Japan, Vol. 3, p. 723, ISIJ, Tokyo (1998).

(13) M. Sorin, V. M. Brodyansky, and J. Paris, "Proceedings of the Florence World Energy Research Symposium, Florence '94, pp. 941-949(1994): Energy Conversion and Management, Vol. 39, pp. 1863-1868(1998).

(14) P. Grassmann, Chem. Ing. Tech., Vol. 22, No. 4, pp. 77-80(1950).

(15) M. Ishida, "Netsurikigaku (Thermodynamics, Principle and Application)" in Creative Chemical Engineering Course 7, Baifuukan Pub., Tokyo (1997).

(16) S. Machida, T. Akiyama, and J. Yagi, Kagaku Kogaku Ronbunshuu (Collected Papers on Chemical Engineering), Vol. 24, pp. 462-469(1998).

# INDEX

www.ingramcontent.com/pod-product-compliance
Ingram Content Group UK Ltd.
Pitfield, Milton Keynes, MK11 3LW, UK
UKHW020821290926
15470UKWH00003B/1295

9 780444 516459